# THE WHICH? GUIDE TO BUYING AND STORING FOOD

ANNE SHEASBY

WHICH?
BOOKS

CONSUMERS' ASSOCIATION

Which? Books are commissioned and researched by
Consumers' Association and published by
Which? Ltd, 2 Marylebone Road, London NW1 4DF

Distributed by The Penguin Group:
Penguin Books Ltd, 27 Wrights Lane, London W8 5TZ

Chapter 3 is by Dr Mike Rayner of the Department of Public Health
and Primary Care, Oxford University

Chapter 4 is by Keith Richards, Senior Lawyer at Consumers'
Association

Typographic design by Paul Saunders
Cover design by Ridgeway Associates
Cover photograph by ACE
Illustrations by Stuart McLean

First edition June 1996

Copyright © 1996 Which? Ltd

*British Library Cataloguing in Publication Data*
A catalogue record for this book is available from the British Library

ISBN 0 85202 640 8

For a full list of Which? books, please write to Which? Books,
Castlemead, Gascoyne Way, Hertford X, SG14 1LH

Typeset by Saxon Graphics Ltd, Derby
Printed in Great Britain by Clays Ltd, Bungay, Suffolk

# CONTENTS

# Introduction

FOOD IS vital for life. It provides us with energy and enables us to grow, to function efficiently and to enjoy optimum health and fitness. What we eat affects our general well-being and thus has a big impact on our lives.

The vast range of foods available to consumers today would have been unimaginable even a generation ago, and it continues to increase. Processing techniques have become extremely sophisticated, giving many foods a longer shelf-life than would once have been possible, while air-freighting has made foods grown in other hemispheres available in less time it would once have taken a farmer in Britain to get his eggs or his vegetables to the local market.

Supermarkets today routinely offer about 20,000 different food lines – an embarrassment of riches that can almost overwhelm. How can shoppers make informed choices? We want to be able to recognise freshness and quality, and to know which fresh foods are in season so that we can buy them and eat them when they are at their best. We want to be sure that, having spent our hard-earned cash on good-quality food, we store it hygienically and for no longer than desirable before we consume it. We may want some basic advice on how to prepare, cook and serve a food, or to know how we can use up something fast approaching its 'use by' date.

And what we never want is food poisoning, which in the UK quadrupled between 1985 and 1995 and is still on the increase. The most common cause of food poisoning is campylobacter, a bacterium which was responsible for 48,000 cases in 1994. New strains such as VTEC (a form of *Escherichia coli*) have emerged which are especially virulent, while our old friends listeria and salmonella are still with us despite a fall-off in their publicity. An improvement in food safety controls throughout the food chain from farm to table is the only way to stamp out food poisoning completely. But those of us who buy food can do a great deal to prevent contamination by storing, handling and cooking it properly.

Most food has a limited storage life and all food deteriorates if kept for too long, becoming less palatable, less nutritious and potentially a threat to health. However, the greatest danger is from cross-contamination, and particularly where storage space for food is limited it is well worth giving some thought to the quantities of food we buy at a time so that we can be sure to store it correctly and never to allow cooked and raw food (the latter more likely to harbour harmful bacteria) to co-habit in the refrigerator.

This book explains why foods need to be selected, stored and prepared with care and how to avoid problems, even in the warmest weather. Its extensive A-Z of Food Storage (Chapter 2) shows at a glance how to handle individual foods, from the everyday basics with which we are all familiar to the exotic items which we may have seen on sale but never tried before.

In addition, the book covers food hygiene and safety, cooking and reheating, the general principles of food storage, including freezing, manufacturers' preservation techniques, food labelling and how to pursue a serious food complaint. The appendices include fruit and vegetable calendars, guidance on weights, measures and egg sizes, meat cuts and cooking methods, herbs and spices, plus essential advice on food poisoning.

# CHAPTER 1

# ENJOYING FOOD AT ITS BEST

THIS CHAPTER offers guidance on food in general, including some of the reasons why certain foods are best treated in certain ways – not just in order to avoid food poisoning, but to get the best value from the money we have spent and to enjoy the food while it is at its best.

## Shopping for food

Before you go shopping, think about what you need for the days or week ahead, and make a shopping list. If you are going to be shopping in only one or two particular stores and you know the layout, list items in the order in which you go round the store.

### Shopping economically

Look at the offers on bulk or multiple purchases: are they foods you would have bought anyway, and would definitely use? Are the goods perishable? It is a false economy to buy something in bulk if you cannot use it up before it goes off. And have you room to store the food as it should be stored (cupboard, refrigerator, freezer or wherever)?

Save money by buying supermarket own-brands.

When trying a new food, buy it in a small quantity first time round, in case you do not like it.

The range of food lines now stocked by most supermarkets runs into thousands, but do not ignore the independent specialist shops, including the delicatessens, for more unusual items, difficult-to-find ingredients, or simply a change from what you regularly buy at the supermarket. However, you may find yourself paying a little more in the specialist outlets.

## Shopping for health

Choose fish, lean meat and poultry more often than you opt for foods that have been processed in some way. This will also give you better value, £ for £.

Where there is a choice, as there is for milk, cheese and spreads, select skimmed or semi-skimmed milk, low- or reduced-fat cheese and spreads. Buy foods such as biscuits, cakes and pastries not routinely but occasionally, as a treat. Choose pasta, rice, bread (wholegrain if possible) and potatoes in place of chips and battered food. Buy lots of fresh vegetables and fruit. Choose low-fat or reduced-fat snacks in place of high-fat crisps and other nibbles.

## Food safety in shops

Retail outlets selling food must by law (Food Safety Act 1990) sell only food that is safe to eat. Local authority Environmental Health Officers are obliged to inspect shops regularly and to prosecute offenders. Food retailers must adhere to strict hygiene standards, in accordance with food safety regulations. This is especially important in outlets where staff handle unwrapped food such as meat, poultry, fish or cheese.

Staff who handle food should wear clean protective clothing and have their hair covered. Raw and cooked meat should be sold from separate chilled cabinets and separate utensils, clingwrap (plastic film) or gloves used. Gloves should be used when handling money and hands washed between handling raw and cooked foods. Needless to say, staff handling food should not smoke while on duty, nor should they work on food counters if they have a cold.

Chilled or frozen products must be displayed below the load line of the chilled or freezer cabinets. Avoid buying foods from over-filled or disorganised cabinets. Avoid buying chilled food which does not feel very cold or frozen food that feels soft. Any surfaces with which food may come into contact should be washed down regularly with a bactericide or detergent, using a clean, disposable cloth.

If you notice what appear to be less-than-hygienic conditions or practices – or staff – in a food store, complain to your local Environmental Health Department at your local authority.

Never buy perishable food which is past its 'use by' date (see pages 32–3 and Chapter 3) or food in damaged packaging, even if it is reduced in price for a quick sale. (Food can legally be sold past its 'best before' date but its quality may have suffered.)

Damaged cans are often sold off cheap, but again, beware. If the damage is to the seams of the can the food inside may have become contaminated by air and germs and the can should not be on sale at all.

### Getting food home in good condition

When doing a major shop, select chilled and frozen foods last, so that they stay as cold as possible for as long as possible. Keep them together in the basket or trolley and when you pack them after passing through the checkout, so that they keep each other cold. In warm weather, particularly if your shopping trip is going to take a while, it is worth transporting chilled or frozen foods in an insulated cool bag or box.

When you pack your shopping prior to taking it home, place the heavier items underneath the lighter, more fragile items to prevent damage. Pack raw and ready-to-eat foods separately, and if you have bought any foods that are likely to drip, pack them in separate bags away from other foods.

Get perishable chilled and frozen foods home as soon as possible after purchase and refrigerate or freeze them immediately, as appropriate.

On arriving home, unpack your shopping and put it away immediately. If loose produce such as fruit and vegetables has roots and soil attached, store it away from other foods to avoid cross-contamination as the soil may carry harmful organisms.

## Food safety and hygiene

High standards of hygiene will help you to prevent food becoming contaminated and hence a potential cause of food poisoning. By following a few simple rules when buying, handling and storing foods, you can minimise the risk of food poisoning in your home and be sure of enjoying food at its best.

## Kitchen and personal hygiene

Keep the kitchen clean and tidy, and kitchen cupboards clean, tidy, dry and as cool as possible.

Never smoke in the kitchen, particularly when handling or preparing food.

Keep work surfaces, sinks, walls and floors, kitchen equipment and utensils clean and dry. Disinfect them often and wipe up spillages as they occur. Keep pets out of the kitchen as far as possible, and never let them walk on the work surfaces.

If you are ill with, for example, a stomach upset or cold, avoid handling food altogether. Cover any cuts with a sterile waterproof dressing and renew it daily until the wound has healed.

Before handling food, remove any jewellery, such as rings and watches, and wash hands in hot, soapy water, making sure your nails are clean. Wash hands again after handling raw food, after using the toilet, and after touching waste-bins or pets.

Keep waste-bins covered and use disposable bin-liners. Wash and dry the bin each time before putting in a new bin-liner. Avoid sorting out soiled domestic washing in the kitchen, particularly on work surfaces which you use for preparing food. Wash your hands thoroughly after handling soiled washing.

When preparing food, wear a clean apron and either roll up long sleeves or wear sleeve protectors. Tie long hair back, away from your face.

Wash work surfaces, and utensils such as knives, between each stage. In particular, never use the same utensil for both raw and cooked food unless you have washed it in between. Wipe the tops of cans before opening them to remove any dust and dirt.

Use a spoon to taste food while cooking – do not just dip your finger into the food. Use the spoon once only rather than dipping it in repeatedly.

Wash dishes and cutlery in hot water with detergent. Rinse them thoroughly and allow to drip dry if possible, instead of using a tea-towel. Wash and dry pet dishes separately from other dishes. Change and wash dish cloths, tea-towels and hand towels regularly – ideally, every day.

Scrub chopping boards, whether wood or plastic, after use with hot water and detergent.

## Storing and cooking

For packaged food, follow the manufacturers' guidelines on how and where to store and cook it.

For food bought loose or raw, use the A-Z of Food Storage (Chapter 2 of this book) to check out the best way to store food, and for how long it is safe to do so. Food that is not stored correctly will not only taste less good, but its shelf-life will be reduced dramatically.

Wash fresh fruit and vegetables under running cold water before use. Be particularly thorough with vegetables that have been in contact with the soil.

Throw away any mouldy, damaged, bruised or discoloured food, and do not eat food which has an 'off' or uncharacteristic bitter smell or taste. If in doubt, throw it away.

If a cooking time is stated on food packaging, or in a cookery book for a recipe you are using, do adhere to it, and remember that for oven cooking you must let the oven reach the temperature in question before you put the food in and start timing.

The centre of hot cooked food should reach an internal temperature of at least 70°C (158°F) for a period of at least 2 minutes. This will ensure that most harmful bacteria are destroyed. The internal temperature of joints of meat or poultry can be checked using a meat thermometer (these are available from supermarkets and department stores). If you are cooking several dishes in the oven at the same time, the food will take longer to cook than if you were cooking only one.

Once food has been cooked, serve and eat it quickly or, if it is not for immediate consumption, cool it as quickly as possible, cover it and store it in the refrigerator.

Use leftover food preferably within one day and always within two days. Cool cooked leftovers as rapidly as possible and chill in the refrigerator as soon as possible. Cool hot food quickly and chill within 1 hour if possible.

Do not re-use food wrappings such as polythene bags, foil or plastic film. Use clean wrapping.

## Reheating

Cooked leftover food should be cooled rapidly and, once cool, placed in the refrigerator (wrapped, or in a covered container) as soon as possible, ideally within an hour. (Do not put food in the fridge while it is still warm, as it will raise the fridge's overall temperature.) The leftover food should be used up ideally within one day and certainly within two days.

Reheat previously cooked food slowly so that it is piping hot throughout, to make sure that any bacteria are destroyed without the outside being over-cooked. If necessary, err on the side of over-cooking rather than under-cooking when reheating.

You may need to add liquid, such as stock, gravy or sauce, to the food you are reheating, depending on how much moisture it lost during initial cooking. Pastries, pies and pizzas on the other hand can become soggy when reheated: these are best warmed up in a conventional oven (as opposed to a microwave). Others, such as cooked fish or meat in a sauce, curries, pasta sauces and so on can be simmered on the hob. Items such as burgers, croquettes, pancakes, stir-fries and so on can be reheated in a frying-pan or wok. Vegetables, fish, puddings and sauces can be steamed, while casseroles and stews can be reheated in a conventional or microwave oven, or on the hob.

To reheat food in a microwave oven, follow the manufacturer's instructions for settings and timings and ensure that the food is piping hot throughout; observe also any recommended standing times.

Frying or deep-frying is suitable for reheating such foods as fritters, croquettes, burgers or chips. Breads and puddings can be reheated in a conventional or microwave oven. Grilled foods such as sausages can be reheated under the grill, over a barbecue, or in a conventional or microwave oven.

If using a grill or barbecue for reheating, or reheating baked foods in the oven, take care not to let the food dry out too much.

Slow cookers can be used for reheating casseroles or stews. Pressure cookers are well suited to reheating items such as Christmas puddings.

Never reheat any type of food more than once. After one reheating, throw any leftovers away.

Serve reheated food promptly. Do not keep it warm as this will encourage bacterial growth and could result in food poisoning.

If you are adding any other ingredients, such as vegetables, to leftovers, cook them first.

Leftover food may need extra flavouring, having lost some flavour since its initial cooking. Check and re-season as necessary before serving. Some leftovers can of course be used as part of a new dish; for example, in pasta sauces, pies, pasties, fish cakes, curries, rissoles, burgers, fritters, croquettes, bubble and squeak, omelettes, (cold) in salads and desserts such as trifle.

# Food poisoning

Many bacteria are harmless and some are even useful – for example, in the production of cheese and yoghurt. Other bacteria can be harmful and may cause food poisoning; these include *Bacillus cereus*, campylobacter, *Clostridium botulinum*, listeria, salmonella and *Staphylococcus aureus*, some of which (such as VTEC) can be fatal.

All types of food poisoning can be prevented. The best way to avoid it is to buy from reputable outlets with a fast stock turnover and make sure that you maintain the highest possible standards of food hygiene in your home wherever food is prepared, cooked or stored.

It is not always easy to recognise a potential source of food poisoning. Sometimes food smells or tastes 'off': if it does, discard it immediately. However, there may be no visual or olfactory warning: food that looks and tastes fine may also contain food poisoning bacteria, which is why good hygiene is so important in reducing the risk.

If you find that a can of food in your store cupboard has become rusty or blown, or is otherwise distorted, throw it away in case the contents have become contaminated.

Symptoms of food poisoning vary depending on the bacteria present but often include nausea, vomiting, diarrhoea and headaches. Some symptoms appear soon after the infected food has been eaten; sometimes, the symptoms may take several days to appear.

If you do become ill through food poisoning, take in plenty of fluids and eat solids once again when you are feeling up to it. If your symptoms are painful or persistent, consult your doctor, particularly if you are in a higher-risk group (these are the elderly, those who are ill or convalescing, babies and young children, and pregnant women).

The higher-risk groups should also avoid eating raw or partly cooked eggs or egg dishes, because raw eggs sometimes contain food poisoning bacteria which are only killed by thorough cooking. If you are making your own mayonnaise, mousse, ice cream or egg drinks at home and want to be sure of avoiding problems, substitute pasteurised eggs for raw ones. If serving ordinary cooked eggs to someone in a higher-risk group, make sure they are thoroughly cooked (for example, serve hard- rather than soft-boiled eggs).

To avoid infection from eating raw or lightly cooked eggs, make sure the eggs used are absolutely fresh and purchased from a reliable source.

Pregnant women and anyone who has a low resistance to infection are also advised to avoid eating soft-ripened cheeses such as brie and camembert, and pâté.

## Food preservation

Raw food is not sterile: it contains bacteria from the surrounding environment. Most bacteria, moulds, yeasts and enzymes are destroyed by heating to a temperature of at least 100°C (212°F). However, some bacteria require higher temperatures to destroy them whilst others cannot be eliminated by heat. In the latter case, food that is already contaminated by heat-resistant bacteria may not be made safe by further heat treatment.

Modern methods of preservation either destroy the bacteria present in food or prevent them from multiplying or growing, thus rendering it safer and producing food with a much longer shelf-life.

The most common ways of preserving foods are canning, pasteurising, sterilising and ultra-heat treatment (UHT), curing, smoking and drying, freezing and refrigeration.

### Canning

Canned foods make convenient store-cupboard standbys, and canning has been a popular method of food preservation for well over a century.

Food to be canned is selected in peak condition and prepared quickly for the canning process, after which it is sealed tightly, cooked and sterilised at high temperatures under pressure in large vessels known as retorts. The temperature and time used for canning foods must be carefully calculated and varies according to the type of food and the size of the can. The cans, most of which are now made from coated steel (lacquered aluminium in the case of drinks cans), are cooled quickly and are then ready for storage at room temperature.

The high heat and pressure ensure that all the food's natural micro-organisms which could cause deterioration are destroyed. The food inside the airtight can is cooked, sterile and stored in a vacuum whilst remaining fresh and nutritious and retaining its texture and flavour. The nutritional value of canned foods is normally high, too, because only top-quality food is worth canning.

Stored correctly, canned food has a long shelf-life, usually of three to five years. Since the EU legislation of 1992, all cans have been stamped with a 'best before' date. Canned food should be stored in a cool, dry cupboard away from heat. When you put new cans away after shopping, put them at the back and bring forward the ones already there so that you use the older cans first.

Once cans are opened, the contents should be treated as fresh food, so if you do not use it all at once place what remains in a covered air-tight container and store in the refrigerator. Food left in opened cans is likely to be contaminated by the metal. Eat the leftover canned food within two days of opening.

Canned food has been cooked and may therefore be eaten straight from the can or heated through before serving, according to the man-ufacturer's instructions.

Canned food and drink must bear a label describing its type, weight and ingredients; nutritional information, storage instructions, serving suggestions and so on may also appear on the label.

## Pasteurising

Pasteurisation is a heat process which destroys many of the bacteria present in food without adversely affecting its appearance or flavour. However, as pasteurisation destroys only the harmful bacteria, the food will not keep for very long because other naturally-occurring bacteria in the food will start to cause decay. Foods which are commonly pasteurised include milk and milk products, eggs and fruit juices.

## Sterilising

Sterilisation is a more severe form of heat preservation which effectively destroys all bacteria naturally present in food, resulting in a product that has a longer shelf-life. The high temperatures used result in a more significant nutritional loss than occurs in a process such as pasteurisation, and the flavour and colour of the product may be altered, as it is, for example, with sterilised milk and cream.

Sterilised foods will keep unopened for several months, sometimes years, if stored correctly. Once opened, they should be treated in the same way as fresh foods – refrigerated and used within a few days of opening.

## Ultra-heat treatment (UHT)

Ultra-heat, or ultra high temperature, treatment is another heat process that destroys all bacteria in food. UHT foods have a long shelf-life: they will keep for several months without refrigeration if stored correctly.

UHT enables food to retain many of its vitamins. Once opened, UHT foods should be treated in the same way as fresh foods – refrigerated and used within a few days of opening. Widely available UHT products include milk and some fruit juices.

## Curing

Curing is a traditional chemical process used to preserve meat and meat products, such as sausages, and to add flavour and colour.

One of several curing agents, such as sodium chloride (common salt), sugar, sodium nitrate, sodium nitrite or vinegar, may be used. The four main methods of application are drying, injecting, adding curing agent directly to the food, and temperature curing.

For dry-curing, salts are rubbed into the meat and the meat is then immersed or pickled in a solution of salts.

Another method is to inject a concentrated solution of curing agents into the arteries, veins or muscular tissue of the meat.

Curing agents can also be added directly to ground meats (such as that used in sausages).

In temperature curing meat is usually cured at a temperature just above 2°C (36°F), depending on the meat.

## Smoking

Smoking is used for preserving meat, fish, sausages and cheese. The process changes the flavour, and often the colour, of the food.

The surface of the food is first impregnated with salt or other preservatives, then it is heated in smoke, often from burning wood, for a period of time ranging from a few hours to several days, depending on the type and quantity of food being smoked. The smoking temperature is in the range of 43-71°C (110-60°F).

Some foods are described as 'hot-smoked' (for example, smoked mackerel), because they cook as they are smoked. 'Cold-smoked' foods, such as smoked salmon, are in effect eaten raw. Some cold-smoked foods such as sausages must be cooked before they are eaten.

## Drying/dehydrating

Micro-organisms and bacteria need moisture in order to grow and reproduce. Its removal prevents the increase of cells. The oldest method is sun-drying. Nowadays, drying – or dehydration – techniques also include spray-drying, roller-drying, freeze-drying and accelerated freeze-drying, and tunnel-drying.

Among the many dried foods available are rice, pasta, pulses, eggs, milk, coffee, tea, meat, fish, herbs, fruit and vegetables. The flavour and texture of some, such as fruit and vegetables, may be changed by the drying process.

Dried foods have a long shelf-life; they will keep for many months in an airtight container in a cool, dry place. If, however, they become damp, the moisture build-up may lead to oxidation, browning and spoiling.

Domestic dehydrators are easy to use and could be worth considering if, for example, you have frequent gluts of fruit or vegetables or want to dry herbs in quantity. The drying period will depend on the type and quantity of the food and may take 4-6 hours or longer.

## Vacuum packaging

Some foods can be temporarily preserved by this method, in which food is wrapped in an impermeable plastic film and air removed under vacuum (see also 'Food packaging', below).

## Freezing

Freezing enables food to be kept for long periods of time while retaining much of its quality, nutritional value and character. Deep-freezing halts deterioration but does not destroy micro-organisms or food poisoning bacteria; however, it renders them inactive, so that while the food is at freezer temperature they are unable to multiply. The recommended temperature for freezing fresh food is –18°C or below. The quality, taste and texture of food which is stored in a freezer even slightly above that temperature is likely to suffer.

For home use the three main types of freezer available are chest freezers, upright freezers and fridge freezers. The size and type you choose depends very much on the space available, the number of people in the household and the level of storage required. Freezers should be kept in a cool, dry place – not next to a heat source such as a cooker, boiler or radiator.

All freezers, including refrigerator ice-boxes, should have a star rating marked on the outside:

* one-star freezer compartments run at a temperature of -6°C (21°F) and should be used only for keeping bought ready-frozen foods, for up to 1 week
** two-star freezer compartments run at –12°C (10°F) and should be used only for keeping bought ready-frozen foods, for up to 1 month
*** three-star freezers run at –18°C (0°F) and should be used only for keeping bought ready-frozen foods, for up to 3 months
**** four-star freezers run at –18°C (0°F) and are suitable for long-term frozen food storage and for freezing fresh food.

Four-star freezers, with the power to take room-temperature items down to –18°C (0°F) rapidly, are the only ones in which you can freeze fresh food yourself.

Most foods are suitable for home-freezing but there are some exceptions, such as bananas, cottage cheese and lettuce (that is, in the main, fresh foods with a very high water content). The A–Z of Food Storage (Chapter 2) has details on freezing specific types of food.

To freeze food at home, follow these guidelines:

- freeze only prime-quality fresh foods, as soon as possible after purchase and certainly on the day of purchase. If food is contaminated before it is frozen it will still be contaminated once it has been defrosted

- handle and pack food hygienically: freezing does not kill bacteria

- freeze or wrap food in suitable containers such as polythene freezer bags, foil and foil containers, freezer wrap, airtight rigid plastic, freezer-proof glass or ceramic containers, or special freezer-to-oven containers

- wrap and pack food carefully: freezer burn (dry, discoloured white patches on the food) or cross-contamination may occur with foods that are not wrapped properly. Proper wrapping also ensures that strong-flavoured foods do not contaminate other foods. Never put unwrapped food in the freezer, unless open-freezing (see page 24)

- exclude as much air as possible from the container or pack – a vacuum pump is ideal for this purpose. Using a drinking straw to suck out the air is unhygienic

- seal all containers well with lids, twist ties or freezer tape

- label and date containers clearly and rotate stock efficiently

- for a large amount of food, turn the freezer to the fast-freeze setting about 6 hours in advance of freezing food and leave on for up to 24 hours after placing non-frozen food in the freezer; for smaller amounts leave the fast-freeze switch on for just a few hours.

The fast-freeze switch works by overriding the freezer thermostat, allowing the temperature of the fast-freeze compartment of the freezer to fall to about –24°C (–12°F), while the rest of the freezer and its contents remain at a temperature of –18°C (0°F). Fast-freezing causes small ice crystals to form rapidly, which keeps the food in better

## Fridge and fridge-freezer compartments

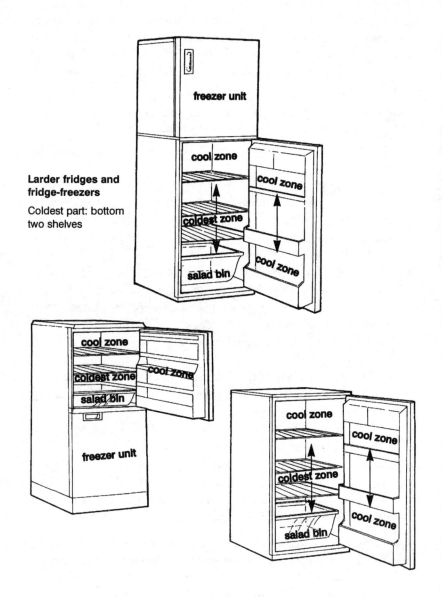

**Larder fridges and fridge-freezers**

Coldest part: bottom two shelves

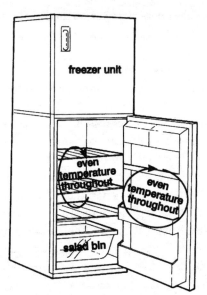

**Frost-free fridges**

Coldest part: as cold air is circulated throughout, temperature varies very little

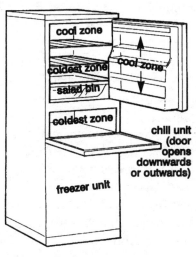

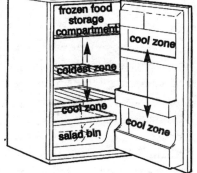

**Three-door fridge freezers**

Coldest part: separate chiller section and bottom shelves of larder unit

**Ice-box fridges**

Coldest part: usually the top and middle shelves

condition than a slower method would. Slow-freezing produces large ice crystals which could adversely affect the texture and taste of the food.

Some foods such as biscuits and small fruits and vegetables (raspberries and peas, for example) can be open-frozen. To open-freeze, spread a single layer of the fresh items on a baking sheet lined with non-stick or greaseproof paper and open-freeze until firm, then transfer them to a freezer bag or box, seal tightly and store in the freezer. Foods frozen in this way are 'free-flowing' – that is, able to be poured out of the pack in whatever quantity you need to defrost.

Get frozen foods home and into the freezer as soon as possible after purchase. Keep frozen foods as cold as possible to prevent them thawing and wrap them well for transporting from the shop to home or, better still, transport them in an insulated cool bag or box.

Do not keep opening the freezer door unnecessarily – warm air will raise the temperature of the freezer. Use a freezer thermometer to check that your freezer is maintaining a steady temperature of –18°C (0°F) and re-check the temperature at regular intervals.

Keep your freezer as full as possible, but not above the load line if it is a chest freezer. Empty spaces will prevent the freezer working as efficiently as it should. Freezers work better and use less energy when full. If yours is not very full, fill the space with basic foods such as bread.

Defrost the freezer when stocks are low. Keep the remaining food frozen, ideally by storing it in another freezer (perhaps a neighbour's). Otherwise, store it in an insulated cool-box or wrap in lots of newspaper or some clean blankets.

Storage times for foods vary. For details see this book's A-Z of Food Storage (Chapter 2), or follow the manufacturer's instructions. Remember that no food should be frozen indefinitely.

Once in a while, defrost and clean the freezer according to the manufacturer's instructions.

In the event of a power cut, keep the door or lid of the freezer tightly shut to prevent warm air entering. If the power cut continues for some time (say, 26 hours for an upright freezer, 30 hours for a chest freezer, provided the door has not been opened and the freezer was reasonably full), wrap the food in several layers of newspaper or clean blankets to reduce the rate of thawing (and hence the deterioration).

**Ready-frozen foods** Store according to the manufacturer's instructions and the freezer's own star-rating system (see above). Try to use frozen foods by the 'best before' date marked on the packaging; and always observe the 'use by' date, after which deterioration could occur.

**Thawing frozen foods** Thawing times of frozen foods vary depending on the thickness and the weight of the food. Thaw frozen food for several hours in the refrigerator or overnight in the refrigerator.

Place frozen food on a plate or tray to defrost and keep it covered, so that any juices that may drip from the food can be collected on the tray or plate and cannot contaminate other foods. Some frozen food may also be defrosted in a microwave oven using the defrost setting – follow the manufacturer's guidelines for timings and ensure that it is defrosted throughout.

Use thawed foods within one day of removing them from the freezer.

Some smaller food items such as peas, sausages, fish fingers and burgers may be cooked from frozen. Follow the manufacturer's guidelines for all ready-frozen foods. In general, larger food items should not be cooked from frozen. If in doubt, defrost the food before cooking.

Never refreeze food once it has thawed unless the food has been cooked in the interim: for example, frozen raw beef which has been thawed and then cooked as part of a casserole can be cooled and frozen.

## Refrigeration

Refrigeration enables perishable foods to be stored for short periods of time. The ideal temperature setting for a fridge is 0-5°C (32-41°F). At this temperature micro-organisms can still grow, but at a slower pace, so food will decay at a much slower rate than it would at, say, room temperature.

The main types of domestic refrigerator are larder fridges, which have no frozen food compartment; fridges with a small ice-box suitable for storing ready-frozen foods and making ice; and fridge-freezers, which comprise a fridge and a freezer, often of equal size and capacity, fixed one on top of the other.

The coldest part of the fridge is at the bottom above the salad drawer, where the temperature should be 0°C (32°F).

Check the temperature of your fridge regularly. If it does not have a built-in thermometer or a thermostatic control (marked in degrees Celsius) you can buy a separate fridge thermometer for about £5 from a supermarket or department store. Place the thermometer in the coldest part of the fridge, ensuring that it is visible as soon as you open the door. Leave the fridge door shut for several hours, preferably overnight, then check the temperature. Adjust the thermostatic control dial and repeat the temperature test procedure, if necessary.

The numbers marked on the thermostatic control dial of a refrigerator are not an indication of temperature, but normally the higher the number on the dial the colder the temperature. Use the dial in accordance with the manufacturer's instructions to adjust the temperature as necessary.

Do not load your fridge to bursting point – cold air must be allowed to circulate for the fridge to work efficiently. If the fridge is overloaded, air is unable to circulate and pockets of warm air may form.

Never leave the fridge door open longer than is necessary, otherwise the internal temperature of the fridge will quickly rise.

Wipe up any spills in the fridge as soon as they occur. Defrost and clean it regularly, in accordance with the manufacturer's instructions. Otherwise, for cleaning use a weak solution of bicarbonate of soda dissolved in warm water (about 5 ml/1 teaspoon bicarbonate of soda to 600 ml/1 pint warm water). Do not use abrasives or fragrant cleaners. Remove and clean each shelf and the inner lining, and wipe dry.

Many fridges now have an auto-defrost facility. If yours does not, defrost it as soon as ice begins to build up. Ice build-up prevents the fridge from working efficiently.

Refrigerator storage times vary from food to food: check the A-Z of Food Storage (Chapter 2) for individual foods.

Never place warm food in the fridge: it will cause the internal temperature of the fridge to rise and increase energy consumption.

**Fridge storage areas**

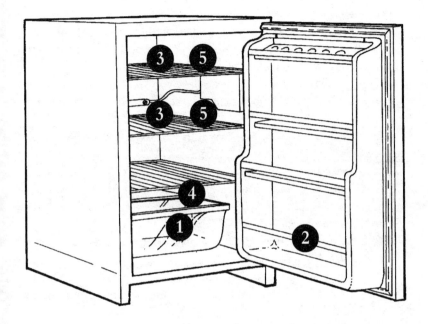

1 Salad drawer: fruit, vegetables and salad items

2 Bottle rack in door: milk and fruit juices

3 Centre and top shelves: butter, margarine, cheese, preserves, salad dressings, spreads, sauces, eggs

4 Shelf above salad drawer: fresh meat, cooked meat, ham, sausages, milk products, cream, fish (store packaged meats separately; store raw meat and poultry below cooked meats and dairy products)

5 Top and centre shelves: convenience foods and cooked items

Cool hot food as quickly and thoroughly as possible before putting it in the fridge by transferring it to a clean, dry, cold container and placing the container in a cool room on a cooling rack to allow air circulation around the food. Cover with a piece of clean muslin or a food cover – do not cover with a lid whilst the food is hot. If possible, divide the food into smaller portions which will cool more quickly.

Food may also be cooled by placing the container in another container of cold water. Change the water often to keep the temperature as low as possible.

Once cooked food has been cooled, refrigerate it as soon as possible.

Use cooked food or leftovers within two days, depending on the type of food (see also 'Reheating', pages 14–15).

Store cooked and raw foods separately in the fridge: cooked foods towards the top and raw foods towards the bottom (the coldest part). For more details, see the A-Z of Food Storage (Chapter 2) and the illustrations on pages 22–3 and 27.

Place raw food such as meat on a plate or tray to catch any drips, even if the food is wrapped.

Always keep refrigerated food covered, or store it in a sealed container. If the food is in its commercial packaging and this is still sealed, leave it like that.

Use chilled foods quickly; never disregard the 'use by' date.

Store high-risk foods such as cook-chill meals in the coldest part of the fridge.

The salad tray is the warmest part of the fridge (about 10-12°C/50-54°F) and should be used only for unprepared fruit, vegetables and salad vegetables.

## Irradiation

Irradiation is a controversial method of treating fresh food that kills unwanted bacteria and pathogens which are responsible for food-borne illnesses. The process destroys insect pests that may be present in grains and other foods which could damage them and cause spoilage while they are being stored.

Irradiation can also delay or arrest the ripening and consequently the moulding process of foods such as strawberries or the sprouting of foods such as potatoes and onions. It also de-activates contaminating bacteria found in foods such as herbs and spices.

In general terms, irradiation extends the storage life of food with minimal effect on its quality and nature, thus reducing spoilage and waste. As for any preservation method, if the food is of poor quality before being processed, it will not be made good by irradiation.

During irradiation, food is exposed to and passed through beams of electrons, x-rays or gamma rays for a specific period of time and is strictly monitored at all times. The food does not come into contact with the radioactive source of radiation and the irradiated food is not radioactive in any way.

Not all foods are suitable for irradiation and with some foods the radiation rays may cause minor chemical changes to the structure. Also, some vitamins, including A and E, are thought to be reduced by irradiation. Foods suitable for irradiation treatment and of which the flavour remains unaffected include herbs and spices, poultry, some seafood such as prawns, and some fruit and vegetables such as strawberries. Foods with a high fat content, for example, dairy products and fatty fish, are not suitable for irradiation as the exposure to the rays accelerates the rancidity process.

If a manufacturer or producer wants to irradiate food it would have to have its premises specially licensed and supervised by the Ministry of Agriculture, Fisheries and Foods (MAFF). In Britain only one company is currently licensed, and the licence is for herbs and spices only. All foods which have been irradiated must be clearly labelled to that effect. Without food labels it is impossible to tell whether food has been irradiated or not, although tests are now being developed that can detect whether a food has been irradiated.

## Organic food

Organic food is grown without artificial or synthetic fertilisers and pesticides and follows a régime of natural systems such as crop rotation which makes the most of natural fertilisers. Animals farmed organically are kept and bred according to high standards of care. If an animal becomes ill, the administration of chemical substances (medicines) is kept to a minimum. Organic farming régimes also require animals to be kept in humane conditions both in transit and at the slaughterhouse. In the words of the UK Register of Organic Food Standards (UKROFS): 'Organic production systems are designed to produce optimum quantities of food of high nutritional quality by using management practices which aim to avoid the use of agro-chemical inputs and which minimise damage to the environment and wildlife.'

The term 'organic' is subject to European Union (EU) and national regulations, by which any establishment wishing to produce, market and sell organic produce must abide. These describe the methods and practices used in organic farming and also govern the processing of organic foods, including the processing aids and ingredients used. Growers, processors and importers of organic food must be EU-registered and are subject to regular inspections of their premises and methods employed. Strict labelling rules also apply. In the UK premises where organic food is produced must be registered with and are regulated and monitored by UKROFS (or a UKROFS-approved body). Inspections of premises are carried out at least once a year.

All organic food on sale has to be clearly marked as such. The word 'organic' on a label is a guarantee that the food has been produced by a registered supplier or producer under the terms of the EU regulations.

## Food packaging

Food is packaged in many different shapes, sizes and materials, including clinging plastic film, foil, greaseproof paper, non-stick baking paper, cardboard and plastic boxes, aluminium, glass and ceramic containers. The primary purpose of packaging is to protect and conserve the contents, in storage and transit. It also serves to make handling and stacking of the product more efficient in the factory, retail outlet and home; it helps to identify the product and can make it look attractive; and it carries essential information (ingredients, weight or volume, storage and cooking instructions, nutritional information, warnings etc.).

All food packaging in Britain must comply with government regulations relating to safety and hygiene. Some flexible plastic film wrapping (referred to generically in this book as 'clingwrap') is not recommended for wrapping fatty foods such as hard cheese and butter, because it is thought that some of the plasticisers it contains could migrate into the food and affect the taste. For this reason, fatty foods such as hard cheese or butter which need to be re-wrapped at home require either another type of wrapping, such as foil or greaseproof paper, or the type of clingwrap which is identified on its own packaging as being suitable for such items.

Clingwrap is suitable for most foods, however, and is especially good for covering bowls and containers as it forms a good seal with the container. It cannot be used in a conventional oven, though, and if it is used in a microwave oven the clingwrap should not be in direct contact with the food, because the plasticisers in it are more likely to migrate into the food at high temperatures if the clingwrap is touching the food.

Some plastic containers such as the bags for boil-in-the-bag products and microwave ready-meals are made of thicker, less flexible material; designed to withstand high temperatures, these containers tend to not to contain plasticisers. Never cook food in plastic containers which are not specifically designed for cooking – for example, plastic mixing bowls or empty yoghurt pots. And before putting a container in a microwave oven, check that the manufacturer has declared it microwave-safe: many food containers are now made specifically for microwave use.

Foil is another good non-permeable wrapping which will adapt to any shape or type of food. It will remain in the shape it is moulded into and provides a good seal over or around food if double-folded at the edge. It can withstand oven temperatures and in cooking may be used to cover or wrap food to ensure that the moisture in the food is retained and the outside of the food is protected from excessive heat or burning. Foil can also withstand freezer temperatures without becoming brittle. Thicker-gauge foil is used to make storage and cooking containers such as pie plates and quiche cases, which are also suitable for freezing.

Greaseproof and non-stick baking paper are other useful wrappings for both hot and cold food. They can withstand lower oven temperatures as well as cold temperatures.

Plastic-lidded, ceramic and glass containers are all good for storing food. Metal containers are suitable for storing some foods but acid foods such as unsweetened stewed rhubarb or sliced citrus fruits should always be stored in a non-metallic container.

Other forms of food packaging used by manufacturers include vacuum packaging and modified-atmosphere packaging. Some foods can be temporarily preserved by vacuum packaging, a process which envelops the food in an impermeable plastic film and then removes the air under a vacuum, thus preventing micro-organisms from entering the food until the seal of the vacuum is broken. Sometimes vacuum-packaged foods have chemical preservatives added to them so that they keep for longer. Foods wrapped in this way usually need to be stored in a cool place such as a refrigerator.

Modified-atmosphere packaging contains a gas or mixture of gases such as nitrogen, carbon dioxide and oxygen rather than air which is present around the food in the package. The air surrounding the food in the package is removed and replaced by a mixture of these gases. This gaseous mixture prevents anaerobic growth in the food – it stops pathogenic spores forming and inhibits bacterial growth. The shelf-life of food packaged in this way is longer than that of fresh food but shorter than that of vacuum-packaged food.

## 'Best before' and 'use by' dates

Most manufactured foods now carry a date stamp of some sort. Such dates must be clearly and prominently stated on the packaging and may not be altered.

The 'best before' date is the one up to and including which the food manufacturer or retailer expects the food to remain in peak condition, provided that it has been stored correctly. After the 'best before' date, the food may still be edible but its appearance, flavour and quality may suffer. The 'best before' date is used for non-perishable foods such as biscuits, pasta, canned foods and rice.

The 'use by' date means that if food is kept and used after this date there could be a risk to health. Food which has passed its 'use by' date should be thrown away unless it has undergone cooking or freezing before the time limit. The 'use by' date is stamped on the packaging of all perishable foods and all food items which could become a safety risk if consumed after the recommended date. Foods that carry a 'use by' date include meats and meat products, poultry, fish and chilled ready-meals, among others. It is illegal to sell food past its 'use by' date.

Foods that do not have to carry a 'best before' or 'use by' date include fresh fruit and vegetables; some fresh cakes and breads (which will have gone stale well before they become a risk to health); fresh produce which is not sold pre-packed, such as meat from a butcher, fish from a fishmonger, delicatessen items, fresh bread sold at bakers, and so on; and foods which keep indefinitely, such as sugar, salt and vinegar. If you are buying non-date-marked foods such as these, you might wish to mark the date on the packaging yourself. Obviously, it makes sense to use foods with the earliest date marks first.

Packaging may also carry date-related recommendations such as 'Best eaten on day of purchase' or 'Eat within 2 days of opening'. Some food products also carry a 'Display until' instruction, aimed at the retailer rather than the consumer.

For more information on food labelling, see Chapter 3.

# CHAPTER 2

# A-Z OF FOOD STORAGE

THE CHAPTER that follows is an alphabetical list, arranged by groups, of specific foods and how to choose, store and use them.
  The food group sections are:

(1)   Baking aids and flavourings
(2)   Beans and pulses
(3)   Biscuits and cookies
(4)   Breads, yeasted goods and cakes
(5)   Crisps and snacks
(6)   Dairy products
(7)   Drinks
(8)   Eggs
(9)   Fats and oils
(10)  Fish
(11)  Fruit
(12)  Grains, cereals and flours
(13)  Meat and poultry
(14)  Nuts
(15)  Pasta and noodles
(16)  Pastry
(17)  Sauces and pickles
(18)  Sugar, preserves and chocolate
(19)  Vegetables.

| Food | How to choose | Storage | Freezing |
|------|---------------|---------|----------|
| **BAKING AIDS AND FLAVOURINGS** | | | |
| Examples: agar-agar, angelica, Bovril, gelatine, Marmite, mixed peel, purées, raising agents, stock cubes or freeze-dried granules, vegetable extract, vinegar, yeast, yeast extract. | Check 'best before' date. Avoid damaged or broken packets and jars. | See label. Generally, store in a cool, dry, dark place. Keeping quality depends on the product. Some products such as tomato purée should be stored in the refrigerator once opened. Store fresh yeast in a polythene bag in the refrigerator for up to 10 days. Use by recommended date. | Most are unsuitable for freezing except as part of a dish. Wrap fresh yeast in foil and freeze for up to 3 months. |
| **BEANS AND PULSES** | | | |
| **Aduki beans** Small, shiny, dark-brown or red beans with a distinctive, tiny white stripe on one side. | Choose plump beans which are bright and clear in colour. Avoid broken, shrivelled or dusty beans. Buy from an outlet which has a rapid turnover. | Store dried beans in a cool, dry place. After opening, store in an airtight container. Use within 1 year. Do not mix new and old beans in storage container. Keep cooked beans in covered container in refrigerator for several days. | Cooked beans freeze well: open-freeze, then pack in small amounts and use as required. Store at −18°C. Thaw before use or add from frozen. When using frozen cooked beans in dishes, reduce the cooking time of the beans slightly so that they remain whole during thawing and reheating. Beans can be frozen in dishes. |
| **Black beans** Medium-sized, shiny, black, kidney-shaped beans with white, slightly sweet flesh. | As aduki beans. | As aduki beans. | As aduki beans. |
| **Black-eyed beans** Small, cream-coloured, kidney-shaped bean with a distinctive black 'eye' on one side (where the bean was joined to the pod). Available whole or split. Pleasant creamy flavour. | As aduki beans. | As aduki beans. | As aduki beans. |

| Preparing and cooking | Typical uses | Alternative forms |
|---|---|---|
| Varies according to the product. Dissolve gelatine in water before combining with other ingredients. Dissolve and activate yeast in warm water before use. Some dried fast-acting yeasts can be mixed directly with other dry ingredients without having been dissolved. | **Agar-agar and gelatine** Stiffen mixtures such as mousses. **Angelica** Decoration. **Bovril, Marmite, vegetable and yeast extracts** Primarily used for flavouring or spreading. **Mixed peel** In cakes such as fruit cakes, puddings. **Purées such as garlic and tomato** Flavouring and colour. **Raising agents** Force baked goods to rise when cooked. **Stock cubes** Flavouring. **Vinegar** Flavouring, pickling. **Yeast** Causes dough (e.g. for bread and buns) to rise when cooked. | |
| Wash and soak in cold water for 4-8 hrs or overnight. Rinse, drain, then simmer gently in a covered saucepan of boiling water for 30-60 mins until tender. Older beans take longer to cook. Add salt at end of cooking time. Drain and use as required. Cooked beans reheat well. | Soups, stews and salads. Ground to a flour and used in cakes, bread and pastry. Also suitable for sprouting. | Canned, frozen. |
| Wash and soak in cold water for 8-12 hrs or overnight. Rinse, drain, then simmer gently in a covered saucepan of boiling water for 50-60 mins until tender. Older beans take longer to cook. Add salt at end of cooking time. Drain and use as required. Cooked beans reheat well. | Soups, stews, casseroles. Popular in Caribbean, Chinese, Mexican and South American cookery. May be used in place of red kidney beans. | Canned, frozen. |
| Wash and soak in cold water for 8-12 hrs or overnight. Rinse, drain, then simmer gently in a covered saucepan of boiling water for 30-45 mins until tender. Older beans take longer to cook. Add salt at end of cooking time. Drain and use as required. Cooked beans reheat well. | Savoury dishes, especially curries and casseroles, Caribbean, Indian and South American dishes. Also suitable for sprouting. | Canned. |

| Food | How to choose | Storage | Freezing |
|------|---------------|---------|----------|
| **Borlotti beans**<br>Plump, kidney-shaped beans which have thin pinkish skin with brownish mottling and a bitter-sweet flavour. | As aduki beans. | As aduki beans. | As aduki beans. |
| **Broad beans**<br>Cream or pale-brown, oval-shaped beans with a fine, floury texture. | As aduki beans. | As aduki beans. | As aduki beans. |
| **Brown beans**<br>Plump, brown, well-flavoured variety of the common haricot bean. | As aduki beans. | As aduki beans. | As aduki beans. |
| **Butter beans**<br>Large, flat, pale, creamy-coloured, oval-shaped beans. Soft, floury texture with a sweetish flavour. | As aduki beans. | As aduki beans. | As aduki beans. |
| **Cannellini beans**<br>Elongated, slender/oval or kidney-shaped creamy white beans with firm, fluffy texture. A type of haricot bean. | As aduki beans. | As aduki beans. | As aduki beans. |
| **Chickpeas**<br>Pale golden or light-brown, hard knobbly peas. Rich, nutty flavour when cooked. | Choose plump peas which are bright and clear in colour. Avoid broken, shrivelled or dusty peas. Buy from an outlet which has a rapid turnover. | Store dried peas in a cool, dry place. After opening, store in an airtight container. Use within 1 year. Do not mix new and old peas in storage container. Keep cooked peas in covered container in refrigerator for several days. | Cooked peas freeze well: open-freeze, then pack in small amounts and use as required. Store at −18°C or below. Thaw before using or add from frozen. When using frozen cooked peas in dishes, reduce the cooking time of the peas slightly so that they remain whole during thawing and reheating. Can be frozen in dishes. |

| Preparing and cooking | Typical uses | Alternative forms |
|---|---|---|
| Wash and soak in cold water for 8-12 hrs or overnight. Rinse, drain, then simmer gently in a covered saucepan of boiling water for 40-60 mins until tender. Older beans take longer to cook. Add salt at end of cooking time. Drain and use as required. Cooked beans reheat well. | Soups, mixed bean salads, Italian dishes, dips, casseroles, stews. | Canned, frozen. |
| Wash and soak in cold water for 8-12 hrs or overnight. Rinse, drain, then simmer gently in a covered saucepan of boiling water for 1-1½ hrs until tender. Older beans take longer to cook. Add salt at end of cooking time. Drain and use as required. Cooked beans reheat well. | Soups, stews, salads, purées. | Fresh, canned, frozen. |
| As broad beans. | Soups, stews, bacon or ham recipes. | |
| Wash and soak in plenty of cold water for 8-12 hrs or overnight. Rinse, drain, then simmer gently in a covered saucepan of boiling water for 1-1¼ hours until tender. Older beans take longer to cook. Add salt at end of cooking time. Drain and use as required. Cooked beans reheat well. | Soups, meaty stews, savoury dishes. | Canned. |
| As borlotti beans. | Good all-purpose bean. Soups, casseroles, salads, Italian cooking. Good substitute for haricot bean. | Canned, frozen. |
| Wash and soak in cold water for 8-12 hrs or overnight. Rinse, drain, then simmer gently in a covered saucepan of boiling water for 1-2 hrs until tender. Older peas take longer to cook. Add salt at end of cooking time. Drain and use as required. Cooked chickpeas reheat well. | Hummus, casseroles, stews, soups. Also suitable for sprouting. | Canned, ground into chickpea flour. |

| Food | How to choose | Storage | Freezing |
|---|---|---|---|
| **Flageolet beans**<br>Small, pale-green or white kidney-shaped beans with a distinctive fresh, delicate flavour. A type of haricot bean. | As aduki beans. | As aduki beans. | As aduki beans. |
| **Haricot beans**<br>Small, greyish-white, plump oval beans with pleasant flavour and texture. Available in a number of varieties. | As aduki beans. | As aduki beans. | As aduki beans. |
| **Lentils**<br>Many different kinds and colours: red, yellow, orange, green and brown. All have similar flattish round shape. Sold either whole or split. The best are the tiny black-green Puy lentils, renowned for their excellent flavour and texture. | Choose lentils which are bright and clear in colour. Avoid broken, shrivelled or dusty lentils. Buy from an outlet which has a rapid turnover. | Store dried lentils in a cool, dry place. After opening, store in an airtight container. Use within 1 year. Do not mix new and old lentils in storage container. Cooked lentils keep in covered container in refrigerator for several days. | Cooked lentils freeze well: open-freeze then pack in small amounts and use as required. Store at −18°C or below. Thaw before using or add from frozen. When using frozen cooked lentils in dishes, reduce the cooking time of the lentils slightly so they remain whole during thawing and reheating. Can be frozen in dishes. |
| **Lima beans**<br>Two varieties and sizes: kidney-shaped creamy white or pale-green beans. | As aduki beans. | As aduki beans. | As aduki beans. |
| **Mung beans**<br>Small, hard, olive-green beans with yellowish flesh. Sweet and creamy in texture when cooked. Available whole, split or skinless. | As aduki beans. | As aduki beans. | As aduki beans. |
| **Peas**<br>Green and yellow, whole and split varieties available. | Choose plump peas which are bright and clear in colour. Avoid broken, shrivelled or dusty peas. Buy from an outlet which has a rapid turnover. | Store dried peas in a cool, dry place. After opening, store in an airtight container. Use within 1 year. Do not mix new and old peas in storage container. Keep cooked peas in covered container in refrigerator for several days. | Cooked peas freeze well: open-freeze, then pack in small amounts and use as required. Store at -18°C or below. Thaw before using or add from frozen. When using frozen cooked peas in dishes, reduce the cooking time of the peas slightly so they remain whole during thawing and reheating. Can be frozen in dishes. |

| Preparing and cooking | Typical uses | Alternative forms |
|---|---|---|
| Wash and soak in cold water for 3-4 hrs. Rinse, drain, then simmer gently in a covered saucepan of boiling water for 1-1½ hrs until tender. Older beans take longer to cook. Add salt at end of cooking time. Drain and use as required. Cooked beans reheat well. | Mixed bean salads. As a vegetable simply served tossed with butter or oil. Also suitable for sprouting. | Canned, frozen, fresh. |
| As flageolet beans. | Cassoulets, soups, bean salads, purées, baked beans, Boston baked beans. | Fresh, canned in tomato sauce as baked beans. |
| Lentils do not need soaking. Simmer gently in a covered saucepan of boiling water for 30-60 mins until tender. Older lentils take longer to cook. Add salt at end of cooking time. Drain and use as required. Brown and green lentils hold their shape when cooked. Red, orange and yellow lentils cook quickly to a purée. Cooked lentils reheat well. | Soups, salads, Indian dishes (as dal or dahl). As a vegetable. | Canned. |
| As flageolet beans. | Soups, casseroles, salads. As a hot vegetable. | Fresh, frozen, canned. |
| Wash and soak in cold water for 3-4 hrs. Rinse, drain, then simmer gently in a covered saucepan of boiling water for 45-60 mins until tender. Older beans take longer to cook. Add salt at end of cooking time. Drain and use as required. Cooked beans reheat well. | Stews, curries, soups, purées, stuffings. Also suitable for sprouting. | Ground to make flour. Sprouted, i.e. beansprouts. |
| Wash and soak in cold water for 1-2 hrs. Rinse, drain, then simmer gently in a covered saucepan of boiling water for 45-60 mins until tender. Older peas take longer to cook. Add salt at end of cooking. Drain and use as required. Cooked peas reheat well. | Soups, pease pudding, mushy peas. | Fresh, frozen, canned. |

| Food | How to choose | Storage | Freezing |
|---|---|---|---|
| **Pinto beans** Kidney-shaped beans with orange-pink mottled skin which becomes pink when cooked. Original ingredient of Mexican re-fried beans. | As aduki beans. | As aduki beans. | As aduki beans. |
| **Red kidney beans** Medium-sized, dark red-brown or deep burgundy-coloured kidney shaped beans with creamy white flesh. Well-flavoured, popular bean which keeps its flavour, shape and colour when cooked. | As aduki beans. | As aduki beans. | As aduki beans. |
| **Soya beans** Several varieties and colours available including yellow, green, brown and black. Bland-tasting, these beans readily absorb the flavours of stronger foods. Most nutritious of all pulses. | As aduki beans. | As aduki beans. | As aduki beans. |
| **Tofu (soya bean curd)** Natural soft and cheese-like bean curd made from soya beans and water. Bland-tasting, tofu readily absorbs flavours from other foods. Available in soft, firm, silken and smoked forms. | Sold chilled in vacuum or modified atmosphere packs. Buy from refrigerated cabinet, check 'use by' date. | Store in refrigerator at 0-5°C. **Unopened** Store in original pack. **Opened** Store original tofu in a bowl of cold water for up to 4 days (changing water daily), marinated and smoked tofu in a covered container for up to 2 days. Change water daily. Tofu does not absorb other flavours unless in direct contact with them. Use by recommended date. | Can be frozen at −18°C or below for up to 1 month. Freeze in original packaging. Becomes noticeably firmer and more fibrous in texture if frozen. Defrost for a few hours at room temperature. |

| Preparing and cooking | Typical uses | Alternative forms |
|---|---|---|
| As flageolet beans. | Mexican re-fried beans, stews, casseroles, soups. A substitute for red kidney beans. | |
| Wash and soak in cold water for 3-4 hrs. Drain and cook rapidly in boiling water for 15 mins, then simmer gently for 1½-2 hrs until tender. Older beans take longer to cook. Add salt at end of cooking time. Drain and use as required. Cooked beans reheat well. | Mixed bean salads, stews, chilli con carne. | Canned, frozen. |
| Wash and soak in cold water for 5-8 hrs. Rinse, drain, then simmer gently in a covered saucepan of boiling water for 3-4 hrs until tender. Older beans take longer to cook. Add salt at end of cooking time. Drain and use as required. Cooked beans reheat well. | Soups, casseroles, purées. As a vegetable, salads. Also suitable for sprouting. | Flour, milk (plain, flavoured, calcium-enriched, no added sugar or salt), basis of textured vegetable protein (TVP – meat substitute). Available as mince or in chunks, bean curd (also known as tofu, q.v.), soy or soya sauce, miso (fermented soya bean paste used in Japanese dishes), noodles, soya dessert, long-life soya products such as yoghurt, soya bean flakes, soya ice. |
| May be used straight from the packet. Rinse in cold water before use. Use raw or fried, grilled, casseroled, barbecued, boiled, marinated, baked or steamed. | Eat on its own or use in soups, sauces, dips, dressings, savoury dishes in place of meat, curries, stir-fries, stews, casseroles, pâtés, desserts. | Smoked, marinated, fermented, frozen, deep-fried, canned, braised. Ready-prepared tofu products include sausages, burgers, tofulomi (vegetarian salami), pâté, 'mayonnaise'. |

| Food | How to choose | Storage | Freezing |
|------|---------------|---------|----------|

## BISCUITS AND COOKIES

| | | | |
|------|---------------|---------|----------|
| Sweet and savoury varieties including chocolate, crackers, oatcakes, semi-sweet, wafer. | Check 'best before' date. Avoid split, open or crushed packets. | If stored correctly, biscuits will keep fresh for weeks. **Unopened** Store in a cool, dark, dry place away from strong light. **Opened** Store as above in an airtight container. Use by recommended date. Do not store biscuits in same container as cakes: they lose their crispness. A few grains of rice placed in the biscuit tin will absorb moisture and keep biscuits crisp. Some tins have a small container attached to the lid containing a substance which absorbs moisture to help keep biscuits crisp. For maximum efficiency, occasionally place this container in a warm oven to dry out. | Cooked home-made biscuits may be frozen. Wrap carefully, seal in foil or freezer bags and freeze at −18°C or below. Defrost before eating, crisp up in oven if liked. Home-made, uncooked biscuit dough may be frozen. Roll up and wrap tightly in foil, then freeze and use within 3 months. Defrost slightly before cooking. Pipe soft mixtures into shapes, open-freeze and pack in freezer bags or boxes when firm. Allow extra cooking time from frozen. |

## BREADS, YEASTED GOODS AND CAKES

**Breads**

| | | | |
|------|---------------|---------|----------|
| Three basic types: white, brown, wholemeal. Many other types including chapati, ciabatta, currant, French, fruit, granary, high-fibre, Hovis, malt, milk, nan, paratha, pitta, poppadom, pumpernickel, rye, soda, softgrain, Vitbe and wheatgerm. | Check 'best before' date. Buy only fresh bread. | Keep in an airtight container in cool, dry place. Store wrapped bread in its original packaging. Do not allow bread to sweat in wrapping or polythene bags. Most wrapped bread will stay fresh for 3-4 days, unwrapped not as long. Some bread is best eaten on day of purchase: see label. Before storing new bread in a bread bin or crock, discard old stale bread and crumbs, wash and dry the bread bin well. Old bread becomes dry and stale and can turn mouldy in a humid atmosphere. Crusty bread will go soft if sealed in plastic. Crisp it up again by heating in the oven for a while. Do not store bread in the refrigerator: this speeds up the staling process. Use by recommended date. | Freeze bread on day of purchase. Overwrap freshly baked bread and packets of bread, pack in polythene freezer bag, expel air, seal and label. Open-freeze buns and delicate breads until firm, then pack in polythene freezer bag. All will freeze at −18°C or below for up to 3 months. Bought part-baked bread and rolls can be frozen in original packaging for up to 4 months. May be cooked from frozen. Bread can be frozen in slices or portions, if desired. Fresh breadcrumbs also freeze well. Thaw bread in wrapper or polythene bag at cool room temperature for several hours, in a microwave oven or slowly in a low oven. Sliced bought bread can be toasted from frozen. |

| Preparing and cooking | Typical uses | Alternative forms |
|---|---|---|
| Some sweet biscuits can be crushed and combined with melted fat or syrup, then used as a cheesecake base. | Eaten on their own, as a snack or treat. Crushed in cheesecake bases. | Packet biscuit mixes, carob, egg- and gluten-free, gluten-free, low-sugar, organic, sugar-free, wheat, wholemeal/wholewheat. |
| To freshen stale bread, moisten and wrap in foil and warm in the oven for about 10 mins. Be careful if using a microwave oven to defrost or heat bread, as it will dry out and harden very quickly. | Sandwiches, snacks, toast, to accompany soups and main-course dishes. As breadcrumbs, bread sauce, bruschetta, croûtons, fried bread, meatloaf, Melba toast, steamed puddings, stuffings. | Bread rolls and buns, bread sauce mix. Chilled products such as garlic bread, dehydrated breadcrumbs. Frozen, home-made, in a can, long-life, organic, packet mixes, partly baked. Quick bread such as soda bread, ready-made, frozen or chilled desserts such as bread-and-butter pudding, summer pudding, scones. Pizza bases are also made from bread dough. |

| Food | How to choose | Storage | Freezing |
|------|---------------|---------|----------|
| **Yeasted goods and cakes** Examples: bagel, brioche, cakes (e.g. cherry, chocolate, fruit, ginger, madeira, sponge), Chelsea bun, croissant, crumpet, Danish pastry, doughnut, eccles cake, hot cross bun, muffin, teacake, waffle. | Check 'best before' date. Buy only fresh yeasted goods and cakes. | **Yeasted goods** (e.g. brioche and croissant) Best eaten as fresh as possible: see label. Store in airtight container or sealed polythene bag in a cool, dry place. **Cakes** Cakes containing fat keep longer than those without. Cover lightly with foil or clingwrap. Some cakes (e.g. gingerbread) are best kept wrapped for a couple of days before eating. Place a piece of fresh apple in the container to keep cake moist, if it is to be stored for any length of time. Use by recommended date. **Iced or filled cakes** Keep best unwrapped in the refrigerator. **Fresh cream cakes** Store in refrigerator, in original packaging or lightly covered with foil or clingwrap. **Rich fruit cakes** Age for at least 1 month before eating. Can keep for up to a year, or more in some cases. | Most are suitable for freezing. Freeze on day of purchase. Wrap and seal in polythene bag or rigid airtight container and freeze at −18°C or below for up to 3 months. Thaw for several hours at cool room temperature. Home-made cakes can be frozen as above; wrap plain cake layers separately. Open-freeze iced cakes and gâteaux until icing has set, then wrap and seal in rigid container. Unwrap before thawing as above. |

## CRISPS AND SNACKS

| | | | |
|------|---------------|---------|----------|
| Examples: breadsticks, crisps, popcorn, pretzels, tortilla chips. Available plain or flavoured. Sold in packets. | Check 'best before' date. Avoid split or open packets. | Store unopened packets in cool, dark place. After opening, seal packet with a clip or rubber band and store in airtight container as above. Will soften and become stale if stored incorrectly after opening. Use by recommended date. | Not applicable. |

## DAIRY PRODUCTS: BUTTERMILK (CULTURED BUTTERMILK)

| | | | |
|------|---------------|---------|----------|
| Cultured milk product with a slightly thick consistency and acid taste. | Buy fresh from refrigerated cabinet. Check 'best before' date. | Refrigerate as soon as possible after purchase. Best kept in original container. Store away from strong-smelling foods in milk rack or dairy section of refrigerator at 0-5°C for up to 5 days. Do not mix old and fresh milk. Use by recommended date. | Not suitable for freezing. |

| Preparing and cooking | Typical uses | Alternative forms |
|---|---|---|
| | Eaten on their own, spread with a fat spread or jam (e.g. scones, croissants), as a snack or treat. Use stale plain cake for trifle. | Packet cake mixes, frozen cakes and gâteaux, partly baked, petit-fours, ready-frozen, small cakes such as brownies, fairy cakes, madeleines, ready-to-bake mixtures sold in cans (e.g. croissants). |
| Not applicable. | Eaten on their own as snack or treat, used for dipping into dips. | Low- or reduced-fat crisps, wholemeal, organic. |
| Buttermilk may be heated or used for baking or in dishes. | Serve chilled as a drink. Use in some savoury breads and cakes such as scones and soda bread, biscuits, salad dressings, confectionery, with fruit. Good substitute for sour milk. | Dried. |

| Food | How to choose | Storage | Freezing |
|------|---------------|---------|----------|

## DAIRY PRODUCTS: CHEESE

**Blue and veined cheeses**

Cheeses marbled with blue veins include Danish blue, dolcelatte, Gorgonzola, Roquefort, Stilton. Creamy, sometimes crumbly texture and often dominant piquant smell.

Check 'best before' date. Choose cheese with firm, not cracked, rind, evenly marbled with blue and creamy inside. Avoid discoloured cheese or cheese with an offensive smell. Poor-quality blue cheese has salty flavour, sharp smell and can be bitter if over-aged. Avoid blue cheese with patchy veining.

Wrap loosely in greaseproof paper or foil and store in airtight container. Store in refrigerator at 0-5°C on top shelf or in dairy section. Unpasteurised blue cheese will keep for up to 1 week, pasteurised for up to 2 weeks. Use by recommended date.

Stilton can be frozen but crust will be soft when thawed. Wrap in foil then seal in freezer bag. Freeze at −18°C or below for up to three months. Thaw slowly in refrigerator or cool room temperature for about 24 hours. Do not refreeze after thawing.

**Fresh cheeses**

Examples: cottage, cream, curd, soft, mascarpone, quark, ricotta. Consistency ranges from smooth to curd, depending on type. Fat content varies. Most have refreshing taste and light texture.

Check 'best before' date. Choose light-coloured cheese which is moist but not sloppy, with a sweet aroma. Avoid cheese seeping moisture or liquid.

Keep tightly covered in container. Store in refrigerator at 0-5°C on top shelf or in dairy section for 2-7 days, depending on type. Use by recommended date.

Cottage cheese in raw state is not suitable for freezing. Freeze other fresh cheeses in carton at −18°C or below for up to 1 month. Some curd cheeses can be frozen for up to 3 months. Thaw overnight in refrigerator. Do not refreeze after thawing.

**Hard and semi-hard cheeses**

Examples: Cheddar, Cheshire, Double Gloucester, Edam, Emmenthal, Gruyère, Jarlsberg, Leicester, Parmesan etc.

Check 'best before' date. Age and price are very much an indication of quality, hence the high cost of an aged cheese such as Parmesan. Look for soft cheese which is moist, not damp, with a nutty aroma, firm cut surface and dry rind. Rinded cheeses are better than rindless or those with an artificial rind. Cheese such as Cheddar should be firm and fresh-looking, not dry or cracked. Look for cheese with uniform colour and texture throughout and no dry patches. Cheese will be soft and bland if not sufficiently mature.

Strong cheese may overwhelm other ingredients. Wrap loosely in greaseproof paper or foil and store away from bland foods in refrigerator on top shelf or in dairy section at 0-5°C for up to 2 weeks. Will turn rancid if stored at too high a temperature. Refrigerate grated cheese in polythene bag for up to 1 week: it will mould quickly if it becomes moist. Use by recommended date.

Hard cheese (e.g. Cheddar) and grated hard cheeses (e.g. Cheddar and Parmesan) can be frozen. Wrap in foil, then seal in freezer bag and freeze at −18°C or below for up to 3 months. Frozen cheese loses some of its flavour and may develop a crumbly texture. Once thawed, it will deteriorate quickly and is best used for cooking. Do not refreeze after thawing.

| Preparing and cooking | Typical uses | Alternative forms |
|---|---|---|
| Different varieties substitute for each other in recipes. Cut whole blue cheeses such as Stilton flat; never scoop. Eat as is, crumble, slice, or use in cooking. Best served at room temperature. | Salads, dips, pasta sauces, sauces, dressings, cheeseboard. | |
| May be cooked in dishes such as baked cheesecakes or sauces. | Served on their own or in salads, spreads, sauces, dips, dressings, desserts such as cheesecakes, tiramisù. | |
| Serve in wedges on cheeseboard, or sliced, grated or diced. Best allowed to sit at room temperature for at least 1 hr before serving. In cooking can be grilled, melted, baked, microwaved etc. | Very versatile in kitchen and on table. Parmesan and Gruyère are especially good for cooking owing to their concentrated flavour. Excellent cooked/uncooked in numerous dishes including fondue, Welsh rarebit, cheese sauce, gratin dishes, sandwiches, pastry, soufflé etc. | Reduced-fat varieties such as Cheddar, vegetarian and organic varieties. Available pre-grated, processed, as processed slices, triangles, portions, sticks and squeezy spreads. |

49

| Food | How to choose | Storage | Freezing |
|------|---------------|---------|----------|
| **Soft cheeses** Examples: soft and rinded cheeses such as Bel Paese, Brie, Camembert, fontina, Monterey Jack, mozzarella, Port Salut and raclette. Often made with unpasteurised milk. | Check 'best before' date. Choose cheese with firm, not dry rind and a soft centre. Avoid hard, swollen, sunken or bitter-smelling cheese and rinded cheese with chalky centre. **Brie and Camembert** Should have white rind, be creamy and soft to the touch and ooze gently when cut. Are bland and firm if under-ripe, bitter if over-ripe. Avoid cheese with cracked or orange-tinged rind: the flavour will be rank. **Mozzarella** Should be soft, white and moist; if too old it becomes tough. | Wrap loosely in foil or greaseproof paper and store in airtight container in refrigerator on top shelf or in dairy section at 0-5°C. Use by recommended date. **Soft pasteurised** Refrigerate for up to 1 week. **Soft unpasteurised** Refrigerate for up to 3 days. **Mozzarella** Refrigerate in brine for up to 3 days. **Pasteurised Brie** Refrigerate for up to 3 days. **Unpasteurised Brie** Store for 24 hrs in cool place. | Generally not suitable for freezing. |

## DAIRY PRODUCTS: CREAM

| Food | How to choose | Storage | Freezing |
|------|---------------|---------|----------|
| **Aerosol cream** Whipped UHT cream packed into aerosol cans. Foamy rather than whipped appearance. Contains added sugar, stabilisers and propellants to make product flow from can. | Short shelf-life. Check 'best before' date. | Once opened, store in refrigerator at 0-5°C. Use absorbent kitchen paper to wipe the nozzle on the can after use, as any cream remaining may harden and go off. Use by recommended date. | Not suitable for freezing. |
| **Clotted cream** Richest of all creams, golden-yellow with a thick granular texture. Rich, buttery flavour. | Buy fresh from refrigerated cabinet. Check 'best before' date. | Keep cool and refrigerate as soon as possible after purchase. Store in milk rack or dairy section of refrigerator at 0-5°C, unopened for 5 days and opened for 2-3 days. Keep tightly covered, otherwise it will harden. Use by recommended date. | Freeze in its carton and store at −18°C or below. Can be frozen for 1 month but may become buttery after that. Thaw for 24 hrs in refrigerator. Do not refreeze after thawing. |

| Preparing and cooking | Typical uses | Alternative forms |
|---|---|---|
| Serve on a cheeseboard or use in cooking. | Cheeseboard or on their own. **Brie and Camembert** Deep-fried in breadcrumbs. **Mozzarella** Pizzas. | Reduced-fat mozzarella. |
| Collapses easily quite soon after serving. Do not heat. If adding as a swirl on hot drinks, serve immediately. | Shake well before use. Serve chilled. Suitable for topping desserts or drinks. Not suitable for decorating as it does not hold its shape for long. | Half-fat aerosol cream. |
| Not recommended for cooking or heating as it tends to separate. Not suitable for whipping or piping. | Spread on scones with jam or in cakes. Serve with puddings and desserts or fruit. | Ready-frozen. |

| Food | How to choose | Storage | Freezing |
| --- | --- | --- | --- |
| **Double cream**<br>Versatile fresh cream, adds richness and texture to dishes. | As clotted cream. | As clotted cream. | Can be frozen. Half-whip cream before freezing. Freeze in carton or in suitable container (leave headroom for expansion) and store at −18°C or below. Will keep in freezer for up to 2 months. Defrost in refrigerator overnight or at room temperature for several hours. Defrost 300ml (½ pint) whipped cream in microwave on low or defrost setting for 1-2 mins. Leave to stand for 10-15 mins before using. Do not freeze cream after its 'use by' date. Do not refreeze after thawing. |
| **Half cream**<br>Thin, light cream suitable for pouring. | As clotted cream. | As clotted cream. | Not suitable for freezing. |
| **Single cream**<br>Versatile cream suitable for pouring. Homogenised has a slightly thicker consistency. | As clotted cream. | As clotted cream. | Not suitable for freezing on its own as it separates on thawing. Freezes well as part of cooked dishes such as quiche. |
| **Soured cream**<br>Slightly thicker than single cream with a refreshing acidic flavour. | As clotted cream. | As clotted cream. | Not suitable for freezing on its own. Freeze only as part of a cooked dish such as cheesecake. |
| **Sterilised cream**<br>Cream which has been heat-treated and homogenised. Has a distinct caramel flavour. | Available in cans. Check 'best before' date. | Keeps unopened for up to 2 years if stored correctly. After opening, keep leftovers in a covered bowl and refrigerate for 2-3 days. Use by recommended date. | Can be frozen in dishes. |
| **UHT/long-life cream**<br>Ultra-heat-treated cream. Flavour similar that of fresh cream. Available in half, single, double and whipping varieties. | Check 'best before' date. | Store in cool, dry cupboard. Keeps for up to 3 months without refrigeration. After opening, treat as fresh cream: keep for 2-3 days in refrigerator. Use by recommended date. | No need to freeze. |

| Preparing and cooking | Typical uses | Alternative forms |
|---|---|---|
| Can be poured or whipped (be careful not to overwhip as it will turn to butter). Frozen cream which has been thawed tends to thicken very quickly when whipped. Add to hot dishes just before serving. | Pouring or whipping. Excellent for puddings, desserts and on fruit. Use whipped for decorating. Use in soups, sauces, casseroles. | Extra-thick spooning cream (not suitable for whipping or piping). UHT cream, ready-frozen, reduced-fat. |
| Not suitable for whipping. | Ideal for coffee, on cereals, in sauces and dressings. On fruit, for enriching savoury sauces and milk puddings. | UHT. |
| Not suitable for whipping. Add to hot dishes just before serving. Do not boil as cream will curdle. | In coffee, savoury dishes such as soups and casseroles, on cereal or fruit, desserts and puddings, custard and sauces. | UHT, ready-frozen, reduced-fat, extra thick (ideal for spooning on puddings, desserts and fruit). |
| Not suitable for whipping. Can be incorporated into dishes prior to baking. Add to hot dishes just before serving. Do not boil as cream will curdle. | Dips, cheesecakes, fruit, baked potato toppings, for enriching soups, sauces and casseroles and enhancing flavour of many sweet and savoury dishes. | Ready-flavoured soured cream (e.g. garlic and herbs or onion and chive flavour). |
| Not suitable for whipping. | Pour or spoon over desserts and puddings; use in dips. | |
| Not suitable for whipping. | Useful for pouring or for adding to dishes. | |

| Food | How to choose | Storage | Freezing |
|------|---------------|---------|----------|
| **Whipping cream** Suitable for whipping or pouring. Will at least double in volume if whipped correctly. | As clotted cream. | As clotted cream. | As double cream. |
| **Crème fraîche** Cultured thick cream with a delicate trace of sourness. | As clotted cream. | Keep cool and refrigerate as soon as possible after purchase. Store on top shelf or in dairy section of refrigerator at 0-5°C for 10-14 days. Older cream will thicken and the flavour may intensify slightly. Use by recommended date. | Suitable for freezing. Store at −18°C or below for up to 1 month. Freeze in carton. Thaw for 24 hrs in refrigerator. Do not refreeze after thawing. |
| **Smetana** Also known as smitane or smatana. A medium-thick soured cream, relatively low in fat, made from skimmed milk enriched with cream and treated with a souring culture. | As clotted cream. | Store on middle shelf of refrigerator at 0-5°C for up to 1 week. Use by recommended date. | Does not freeze well. |

### DAIRY PRODUCTS: FROMAGE FRAIS

| Food | How to choose | Storage | Freezing |
|------|---------------|---------|----------|
| **Fromage blanc** A type of fresh, white, soft curd cheese, light and spreadable. Similar to fromage frais but smoother. Sold in specialist cheese shops, delicatessens and some supermarkets. | Buy fresh from refrigerated cabinet. Check 'best before' date. | Refrigerate as soon as possible after purchase. Store in milk rack or dairy section of refrigerator at 0-5°C for up to 5 days. Use by recommended date. | Not suitable for freezing. |
| **Fromage frais** Lightly fermented, soft creamy fresh curd cheese, sometimes enriched with cream, resembling thick yoghurt in texture. Less tangy than yoghurt, its flavour ranges from mildly acidic to creamy. Available in plain and flavoured varieties, low-fat or medium-fat. Fat content varies from less than 1% to about 8%. | As fromage blanc. | Refrigerate as soon as possible after purchase. Store in tightly covered tub in refrigerator at 0-5°C. Storage time depends on fat content. Use by recommended date. | Freezes reasonably well. Store at −18°C or below for up to 1 month. Thaw overnight in refrigerator. May separate slightly. Stir well after thawing. Do not refreeze after thawing. |

| Preparing and cooking | Typical uses | Alternative forms |
|---|---|---|
| Add to hot dishes just before serving. Do not boil as cream will curdle. Frozen cream which has been thawed tends to thicken very quickly when whipped. | Ideal for piping or folding into cold desserts such as mousses to give light texture. Use in cheesecakes, for filling cakes or pastries. Will not hold its shape on hot foods. Also suitable for pouring. | Ready-frozen, UHT, reduced-fat. Ready-whipped cream, which often contains sugar, is available frozen or fresh. |
| Suitable for sweet and savoury cooking. High fat content prevents curdling when heated; can therefore be used for soups, sauces and savoury dishes. Not suitable for whipping. | Gives a slight tang to many sweet and savoury dishes and acts as a good thickener. Use in place of soured cream or cream. Ideal for spooning over puddings and desserts. | Reduced-fat. |
| Add at end of cooking. It will curdle if allowed to boil. | Good low-fat substitute for cream or soured cream. Used in sweet and savoury cooked dishes, particularly those of Eastern Europe and Russia. | |
| Add to dishes at end of cooking time. Do not boil as it curdles easily. Before cooking with it, stabilise by stirring 2.5-5ml (½-1 tsp) cornflour into 150ml (¼ pint) fromage blanc, to reduce risk of curdling. | Serve with fruit in place of cream, on its own or with sugar, in tartlets and sauces. | |
| Add to dishes at end of cooking time. Do not boil as it curdles easily. Before cooking with it, stabilise by stirring 2.5-5ml (½-1 tsp) cornflour into 150ml (¼ pint) fromage frais to reduce risk of curdling. | In sweet and savoury dishes. As a topping in place of cream, stirred into sauces or with fruit. Often used in desserts in place of cream, salad dressings, dips. | |

| Food | How to choose | Storage | Freezing |
|------|---------------|---------|----------|
| **DAIRY PRODUCTS: MILK** | | | |
| **Condensed milk** Sweet, thick, syrupy, rich and creamy milk: an evaporated-milk product containing added sugar made from whole, partly skimmed or skimmed milk. | Check 'best before' date. | Store in cool, dry cupboard. Use by recommended date. | Can be frozen in dishes. |
| **Dried milk** Skimmed or whole-milk granules or powder which are reconstituted by being mixed with water. | Check 'best before' date. | In its sealed airtight container in cool, dry cupboard, will keep for several months to 1 year. After reconstituting, treat as fresh milk. Use by recommended date. | Not applicable. |
| **Evaporated milk** Rich, thick and creamy unsweetened, concentrated sterilised milk with distinct flavour. | Check 'best before' date. | Store in a cool, dry cupboard. After opening, treat and store as fresh milk. Use by recommended date. | Can be frozen in dishes. |
| **Goats' milk** Less creamy taste than whole milk and whiter in colour. Slightly stronger and more tangy than cows' milk. Available fresh and frozen. | Buy fresh from refrigerated cabinet. Check 'best before' date. | Refrigerate as soon as possible after purchase. Store in refrigerator at 0-5°C. Use by recommended date. | Can be frozen for up to 1 month. Do not refreeze after thawing. |
| **Homogenised pasteurised milk** Whole milk processed so that cream is evenly distributed throughout, hence no visible cream line and no shaking required before use. | As whole milk. | As whole milk. | Freeze at −18°C or below in waxed carton or plastic container, not a bottle, for up to 1 month. Thaw slowly in refrigerator. While thawing, gently shake occasionally. Do not refreeze after thawing. |
| **Semi-skimmed pasteurised milk** Partially skimmed milk from which just over half the cream has been removed. Tastes less rich than whole milk. | As whole milk. | As whole milk. | As homogenised milk. |

| Preparing and cooking | Typical uses | Alternative forms |
|---|---|---|
| Serve as is or use in dishes. | For making desserts or as a topping, in hot drinks such as tea, coffee or hot chocolate, in toffee or fudge, poured over fruit. | |
| Can sometimes be mixed with dry ingredients in baking. Otherwise reconstitute with water before use. | In cooking; add to hot drinks such as tea and coffee; or use with dry ingredients in baking. Useful fresh milk substitute. **Not suitable for babies.** | |
| Serve as is or use in dishes. | In sweet and savoury cooking, as a dessert topping. Use as is or thin down with water. May be whipped like fresh double cream. | Reduced-fat or 'light'. |
| Use in place of cows' milk. | As for cows' milk. | Yoghurt. |
| As whole milk. | In baking and in all dishes containing milk, such as drinks, puddings, soups and baking. | |
| As whole milk. | As for whole milk. Use in hot drinks such as tea, coffee or hot chocolate, or for making batters, milk puddings, sauces. | Flavoured milks and thick shakes, fortified, in a can, organic, sterilised, UHT. |

| Food | How to choose | Storage | Freezing |
|---|---|---|---|
| **Sheep's (ewes') milk** Looks and tastes creamier than cows' milk. Very white in colour and slightly sweet, with no distinct flavour. Available fresh or frozen. | As goats' milk. | As goats' milk. | Not suitable for freezing. |
| **Skimmed pasteurised milk** Milk from which almost all the cream and fat have been removed. Looks and tastes less creamy than whole or semi-skimmed milk. | As whole milk. | As whole milk. | As homogenised milk. |
| **Soya milk** Non-dairy milk made from soya beans. Looks creamy and is bland and odourless. Useful for those sensitive to dairy milk. | As goats' milk. | As goats' milk. | Not suitable for freezing. |
| **Sterilised milk** Golden-coloured milk with faint but characteristic 'cooked' caramel flavour and a rich, creamy taste. Available whole, semi-skimmed and skimmed. | As whole milk. | Unopened, keeps for several months in a cool place. After opening, treat as fresh pasteurised milk and refrigerate at 0-5°C for up to 5 days. Use by recommended date. | As homogenised milk. |
| **UHT milk** Also known as long-life milk. Bland-tasting and packed in sterile, foil-lined airtight cartons. Less suitable for drinking. Best for cooking. Available whole, semi-skimmed and skimmed. | As whole milk. | As sterilised milk. | As homogenised milk. |

| Preparing and cooking | Typical uses | Alternative forms |
|---|---|---|
| As goats' milk. | Use as for cows' milk; ideal for milk puddings. | Yoghurt. |
| As whole milk. | Use as for whole milk, in hot drinks, soups, making sauces, milk puddings, cakes. **Not suitable for babies.** | Dried, flavoured milks and thick shakes, fortified, sterilised, UHT. |
| Tendency to curdle in very hot drinks such as tea. | Use in place of dairy milk in hot and cold drinks, custards, milk puddings, sauces etc. | Sweetened with sugar, honey or apple juice; flavoured with chocolate, banana, strawberry. |
| As whole milk. | In coffee, custards, milk puddings, other sweet and savoury dishes. | |
| As whole milk. | Best used in cooking. | Organic soya milk, lactose-reduced, soya, flavoured. |

| Food | How to choose | Storage | Freezing |
|------|---------------|---------|----------|
| **Whole pasteurised milk** Standard milk with noticeable layer of cream on top. Most popular milk. | Buy from a refrigerated cabinet. Check 'best before' date. | Refrigerate as soon as possible after purchase (keeps best in the container in which bought). Store away from strong-smelling foods in milk rack or dairy section of refrigerator at 0-5°C for up to 5 days. Do not mix old and fresh milk. Use by recommended date. | Not suitable for freezing, tends to separate when thawed. |

### DAIRY PRODUCTS: YOGHURT

| Food | How to choose | Storage | Freezing |
|------|---------------|---------|----------|
| **Drinking yoghurt** Made from yoghurt with added milk and fruit juice or puréed fruit to give drinking consistency. Available as fresh or UHT product in variety of flavours. | Buy fresh from refrigerated cabinet. Check 'best before' date. | Refrigerate as soon as possible after purchase. Store in milk rack of refrigerator at 0-5°C for up to 14 days. Use by recommended date. | Freeze some in carton or airtight container at −18°C or below for up to 3 months: see label. Thaw overnight in refrigerator. Do not refreeze after thawing. |
| **Greek yoghurt** Made from cows' or ewes' milk. Thick, creamy consistency with 1-11% fat content. Available in plain, fruit and flavoured varieties. | As drinking yoghurt. | Refrigerate as soon as possible after purchase. Store in refrigerator at 0-5°C for up to 14 days, depending on fat content. Use by recommended date. | Not suitable for freezing as it separates on thawing. |
| **Low-fat and natural yoghurt** Versatile product with characteristic slightly acidic, refreshing flavour. Available in plain, fruit or flavoured varieties. | As drinking yoghurt. | As Greek yoghurt. | Freeze at −18°C or below in rigid container: fruit yoghurt for up to 3 months, natural for up to 2 months. Thaw slowly in refrigerator, stir well after thawing. Do not refreeze after thawing. |

### DRINKS

| Food | How to choose | Storage | Freezing |
|------|---------------|---------|----------|
| **Carbonated drinks** Varieties include cola and lemonade. | Check 'best before' date. | Store in cool, dry place. Use by recommended date. | Not suitable. |

| Preparing and cooking | Typical uses | Alternative forms |
|---|---|---|
| Scorches easily so use moderate temperatures. | General-purpose milk used in many sweet and savoury dishes such as batters, breads, custards, milk puddings, ice-cream, in hot and cold drinks, on cereals. | Bio, flavoured milks and thick shakes, gold-top (Channel Islands) milk, lactose-reduced, organic, untreated (unpasteurised) milk, sterilised, UHT. |
| Not suitable for cooking. Do not boil. Shake before drinking. | As a drink, sauce. | UHT. |
| May curdle if boiled. Can be heated only as part of a baked dish. Not suitable for whipping. | Serve with fresh fruit or fruit compotes, honey and nuts; add to soups, sauces. | Reduced-fat. |
| Add to dishes at end of cooking time. Do not boil as it curdles easily. Before cooking with it, stabilise by stirring 2.5-5ml (½-1 tsp) cornflour into 150ml (¼ pint) yoghurt to reduce risk of curdling. | Useful for both sweet and savoury dishes, eaten on its own, served on cereal, fruit desserts, used in place of cream as a dessert topping, in dips, drinks, marinades, salad dressings, soups. | Bio, creamy, custard-style, diet, French-style, goats', Greek, high-fibre, live, organic, reduced-calorie, sheep's, soya, sterilised, stirred and set, UHT, whole, whole-milk, very low-fat, with separate fruit portion or layer. |
| Not suitable. | As drink. In ice-cream sodas. | Reduced-calorie, no added sugar. |

| Food | How to choose | Storage | Freezing |
|------|---------------|---------|----------|
| **Coffee** Produced from coffee beans. Many types of whole beans or ground granula (instant) beans available in a variety of roasts. Flavour and aroma vary from strong and bitter to full and delicate. Character and quality vary enormously. The best instant coffees are freeze-dried. | Buy beans regularly and in small quantities. Check 'best before' date. | Keep indefinitely in airtight container in a cool, dry place. May be stored in freezer. Ground beans lose freshness more quickly than whole beans; the more finely ground, the more rapid the loss of taste, so use as soon as possible or by recommended date. | Coffee beans and vacuum-packed ground coffee freeze well. Seal in freezer bags and freeze at −18°C or below for up to 6 months. |
| **Fruit juice** Liquid extracted from raw or cooked fruit. Many flavours available in various forms from freshly squeezed to long-life. | Check 'best before' date. | Depending on type, store in cool, dry cupboard or in refrigerator at 0-5°C – see label. Use by recommended date. | Can be frozen in dishes. |
| **Squashes, cordials and fruit syrups** Squashes and cordials are fruit syrups or sweetened non-alcoholic fruit drinks. Syrups are clear, unsweetened fruit juices. | Check 'best before' date. | Store in sealed container in a cool, dark place. Syrups keep for up to 1 year. Use by recommended date. | Freeze in small plastic containers, leaving headroom for expansion, or in ice-cube trays. Colour and flavour are preserved when frozen. |
| **Tea** Made from processed leaves of an evergreen bush. Three types: green, black and oolong. Two main grades: whole leaf and broken leaf. Numerous varieties available loose or in bags from many countries including China, Ceylon, India and Japan. | Taste depends on where and how tea is grown. Check 'best before' date. | Store in airtight container in a cool, dark, dry place. Use by recommended date. | Not applicable. |

| Preparing and cooking | Typical uses | Alternative forms |
|---|---|---|
| Ideally, grind fresh beans as you need them. Adjust grain to suit method of making, e.g. finely ground for filter method, coarse-ground for percolator method. Use freshly drawn water, just off the boil. Allow 5-10ml (1-2 tsp) coffee per 150ml (¼ pint) water. For simple coffee essence, mix equal volumes of instant coffee and boiling water, cool before use. To brew coffee, simply add boiling water. Strength – from weak to strong – depends on how much coffee is used and how long it is allowed to brew. If left for too long a bitter flavour will develop. Do not reheat coffee more than once. | As café au lait, espresso, Irish coffee etc. As flavouring in cakes and fillings, custards, granita, ice-cream, icing, mousses, sauces, soufflés. Marries well with chocolate. | Many blends from Brazilian to Kenyan; many roasts from light to continental; caffeinated and decaffeinated, coffee essence; instant cappucino; instant coffee-flavoured liqueurs. |
| Delicious and refreshing served on its own or combined with other flavours such as herbs and spices. Vegetables such as carrots and tomatoes can also be processed for their juice. Squeeze fresh juices at home. | On its own as drink; in punches, desserts and puddings, marinades, savoury dishes, sauces. | Frozen, concentrated, dried. |
| May be diluted and served as a drink or used in other beverages. | Syrups are a versatile ingredient – as they are concentrated a little goes a long way. Use for flavouring drinks, as a sauce for desserts and cakes, in fruit sundaes, ice-cream, ice-cream sodas, fruit salads, fruit cups and shakes, jellies; to enhance mousses; dilute with water, soda, milk, yoghurt; mix with white wine to make sodas. Cordials make refreshing drinks when diluted with chilled soda water, sparkling fruit juice or wine. | Reduced-calorie, no added sugar. |
| Use freshly drawn boiling water. Warm teapot with hot water, empty, add tea, pour on water, stir and leave to stand for 3-5 mins, stir and serve. | As hot beverage, drunk on its own or with milk, sugar or lemon. Used in cooking teacakes and tea loaves, fruit breads such as barmbrack, tea bread, syrups, ice-cream, iced tea, punches, sorbet, light desserts. For steeping dried fruit such as prunes. | Iced, herbal, flavoured with spices, flowers, fruits or essences such as mango, lime, wild cherry, chocolate, mint, jasmine, rose, orange blossom. In a can, instant, instant granules, instant flavoured tea such as lemon or blackcurrant, smoked tea. |

| Food | How to choose | Storage | Freezing |
|---|---|---|---|

## EGGS

**Hens' eggs**
Familiar standard brown or white eggs. Several types available including barn, free-range, deep-litter, laying-cage, perchery. Sold in different grades and sizes/weights.

*How to choose:* Buy from reputable source; avoid dirty or damaged eggs. Look for lion symbol on egg boxes (guarantee of quality, freshness and high standard of hygiene). Check 'best before' date on box or on eggs. Fresh eggs have thick whites which keep yolks firmly in place. The older the egg, the runnier the white.

*Storage:* Store in refrigerator at 0-5°C. Use within 3 weeks of purchase or by recommended date. Eggs are porous and will absorb strong odours. **Eggs in shell** Refrigerate pointed end-down in their box or a rack (freshness deteriorates at warmer temperatures). **Whole eggs out of shell** Cover and refrigerate for up to 2 days. **Yolks** Add a little water, cover and refrigerate for up to 2 days. **Whites** Cover and refrigerate for up to 1 week. **Hard-boiled eggs** Refrigerate in shell for up to 2 days.

*Freezing:* Freeze yolks and whites separately or whole at −18°C or below for up to 3 months. **Whites** Store in freezer container, cover surface with greaseproof paper. **Yolks** Mix 2.5ml (½ tsp) salt or sugar to every 3 yolks and store in freezer container. **Whole eggs** Do not freeze in shell. Lightly beat, mix with 2.5ml (½ tsp) salt or sugar to every 3 eggs and store in freezer container. **Hard-boiled eggs** Do not freeze as they become rubbery. Thaw in refrigerator or cool place. Use on day of thawing. Do not refreeze after thawing. Can be frozen in dishes.

**Other eggs**
Examples: duck, goose, guinea-fowl, gull, partridge, pheasant, pigeon, quail and turkey eggs.

*How to choose:* Availability varies. Buy duck and goose eggs from a reputable source. Avoid dirty or damaged eggs. Check 'best before' date (if stated).

*Storage:* Use by recommended date or as soon as possible after purchase. Store in refrigerator at 0-5°C.

*Freezing:* Can be frozen in dishes.

## FATS AND OILS

**Butter**
A natural dairy product made by beating cream until it thickens and separates. Usually made from cows' milk. Colour ranges from pale to deep yellow, depending on type of cow and its diet. Two basic types: sweet cream and lactic, available salted or unsalted.

*How to choose:* Buy from refrigerated cabinet. Check 'best before' date. Choose cool, firm butter with dry wrapping, fresh, untainted aroma and uniform colour.

*Storage:* Butter picks up other flavours and odours easily. Keep wrapped or in a butter dish as light causes rancidity. Best kept in refrigerator at 0-5°C. **Salted/unsalted butter** Store well wrapped away from strong-smelling food in refrigerator for up to 4 weeks. **Clarified butter** Store as salted butter for up to 3 weeks. Use by recommended date.

*Freezing:* Freeze overwrapped (or place packs of butter in freezer bags and seal tightly) and store at −18°C or below. **Salted/clarified butter** Freeze for up to 3 months. **Unsalted butter** Freeze for up to 6 months. **Flavoured butter** (such as herb butter) Freeze as salted butter. **Butter curls or pats** Open-freeze before layering between greaseproof paper and freezing in a rigid container. Thaw butter in refrigerator overnight. Do not refreeze after thawing.

| Preparing and cooking | Typical uses | Alternative forms |
|---|---|---|
| When handling eggs make sure hands, preparation surfaces, utensils and containers are clean. Do not use cracked or dirty eggs. For best results remove from refrigerator 30 mins before cooking. Eggs over-cook easily so always use a gentle heat, unless making omelettes or similar. | For all types of cooking such as soufflés, quiches, meringues, omelettes, cakes, custards and other baked items. Use to thicken, set, enrich, emulsify, coat, bind or glaze other foods; boiled, baked, scrambled, fried, poached, pickled. | Bantam eggs, similar to hens' eggs but half the size. Pullet eggs, laid by a hen in her first year. Pasteurised, dehydrated, pickled, smoked, cured (1,000-year-old), four-grain eggs. |
| Cook duck eggs well (ideally, serve hard-boiled). | Use duck, goose and turkey eggs in place of hens' eggs, on their own or in dishes. Quail eggs are often served boiled in shell as a garnish. | Salted duck eggs, pickled. |
| May be used hard, soft, melted, creamed etc. Does not spread straight from refrigerator: bring up to room temperature. Heat alters the form and flavour of butter. It contains water and milk solids, so burns at a lower temperature than other cooking fats. Take care when using high heat. Not suitable for deep-fat frying. Clarified butter is better for high-heat cooking methods such as frying. Add oil to butter when frying to prevent burning. | Adds substance, flavour, richness and texture to food. Use alone as a spread and for cooking or seasoned with flavourings such as chilli, garlic, herbs, honey, nuts. Ideal for baked goods, light sautéing, pastries, sauces. | Butter spreads, clarified, concentrated, half-fat, ghee, whey, whipped, May also be made from other milks such as camel, goat, sheep, water buffalo, yak. |

| Food | How to choose | Storage | Freezing |
|------|---------------|---------|----------|
| **Cooking fats (hard and soft)** Examples: dripping, lard, hard or block and soft margarine, shortening, solid oils, suet. | Buy from refrigerated cabinet. Check 'best before' date. Look for fresh, clear colour with no dried edges. | Turns rancid if stored incorrectly or for too long. Keep tightly covered: fats pick up other flavours and aromas easily. Store in refrigerator at 0-5°C for up to 2 months. Refined fats such as lard and grated suet may be stored in packets or covered container for months in refrigerator. Use by recommended date. | Freeze at −18°C or below for up to 3 months. Overwrap packs or tubs in freezer bags and seal. Thaw overnight in refrigerator. Do not refreeze after thawing. |
| **Low-fat spreads** Spreading alternatives to butter and margarine include low-fat spreads, reduced-fat spreads and very low-fat spreads. All spread easily from chilled. | Buy from refrigerated cabinet. Check 'best before' date and nutritional information on pack. | Keep covered in refrigerator at 0-5°C. Use by recommended date or within a certain number of days after opening; see label. | Some are suitable for freezing at −18°C or below for up to 3 months, some do not freeze well – see label. Thaw overnight in the refrigerator. Do not refreeze after thawing. |
| **Oils** Examples: almond, corn, grapeseed, hazelnut, olive, peanut, rapeseed, sesame, soya, sunflower, vegetable, walnut. Each has own specific character, colour, flavour and culinary purpose. | Look for fresh aroma and flavour, not rancid or acid, and clear, bright colour. Taste and colour depends on oil. Check 'best before' date. | Air, heat and light will cause oil to oxidise and turn rancid. Store in an airtight container in a cool, dark cupboard away from sunlight, preferably in a can or dark-coloured bottle. Oils may solidify in cold temperatures or in the refrigerator but will clear at room temperature. After opening, oil will keep for 6-12 months if stored correctly. Use by recommended date. | Not suitable for freezing. |

| Preparing and cooking | Typical uses | Alternative forms |
|---|---|---|
| Hard fats will not spread straight from refrigerator. Most are suitable for shallow-frying and sautéing: margarine is least suitable for frying, especially deep-fat frying. White fats such as lard are best for deep-frying. | Hard margarine, suet and refined cooking fat are best for baking; lard and solid vegetable oils are more versatile. Use for cakes, frying, pastries, puddings, roasting, sauces, spreading. | Vegetable suet, 'light' reduced-fat suet, vegetarian margarine. |
| Use in cooking is limited. See label. Low-fat spreads are not suitable for frying or deep-fat frying. Very low-fat spreads are suitable only for spreading due to high water content. | Spread on bread, toast, scones etc. Some can be used in cooking, e.g. all-in-one sauces and cakes. | Half-fat butter. |
| Oils may be used as they are to add flavour, richness, smoothness and consistency to cooking. Use as fat for browning. Use grapeseed, olive, rapeseed, sesame and sunflower for frying: these withstand high temperatures. Use others such as almond, hazelnut, walnut (i.e. those with a distinct taste) more for flavouring and seasoning. Use vegetable, peanut and corn oil for deep-fat frying. When frying food in oil maintain a constant correct temperature, otherwise food will absorb oil, become soggy and not cook properly. | Uses depend on oil: frying, roasting, baking, marinades, Indian, Chinese and Mediterranean cooking, bruschetta, stir-fries, basting, deep-frying. Flavoured oils (olive, nut-based etc.) are best for salad dressings and for pasta and pasta sauces. | 'Light' oils, spray oils, infused oils (oils infused with herbs, spices or other flavourings, such as truffles), exotic oils such as avocado, mustard and poppy seed, individual sachets of oil. |

| Food | How to choose | Storage | Freezing |
|------|---------------|---------|----------|

## FISH

**White fish**

Three main groups: round, flat and freshwater. In white fish the oil is concentrated in the liver. Examples: bass, cod, coley, haddock, halibut, lemon sole, monkfish, plaice, whiting.

Sold whole and as cutlets, steaks, fillets (skinned and unskinned). Availability varies, depending on region and weather conditions. Look for Seafish Quality Award symbol representing fish retailers who achieve high standards. Buy from refrigerated counters or displays on well-packed ice. Check 'use by' date. **Whole** Look for shiny, moist skin with bright, natural colouring, clear, bright, unsunken eyes without discoloured slime or cloudiness, and firm flesh without abrasions that does not indent under finger pressure. Gills, if present, should be bright red and scales intact. Smell should be sea-fresh. **Fillets** Should be neat, firm, moist, trim, white translucent colour, no discoloration or bruising. **Frozen** Should be hard with no sign of partial thawing or freezer-burn. Packaging should be intact.

Use fresh fish at once or store near bottom of refrigerator at 0-5°C for up to 2 days. To refrigerate, rinse, pat dry with kitchen paper, place on a plate and cover with clingwrap or another plate. Use by recommended date.

Maximum storage period is 6 months. Freeze directly after purchase unless fish has previously been frozen: see label. To freeze, wash fish and pat dry with kitchen paper, double-wrap in polythene, freezer-wrap or foil for extra protection and pack in freezer bags. Expel as much air as possible and store at −18°C or below. Thaw overnight in refrigerator or in microwave using defrost setting. Do not thaw in water. Never refreeze fish after thawing.

**Oily, oil-rich/fatty fish**

In oily/fatty fish, the oil is dispersed throughout the flesh. Examples: anchovy, herring, kipper, mackerel, salmon, sardine, trout, tuna.

Sold whole, canned and as fillets. Check 'use by' date. Look for Seafish Quality Award symbol representing fish retailers who achieve high standards. Buy from refrigerated counters or displays on well-packed ice. Check 'use by' date. **Whole** Look for shiny, moist skin with bright, natural colouring, clear, bright, unsunken eyes without discoloured slime or cloudiness, and firm flesh without abrasions that does not indent under finger pressure. Gills, if present, should be bright red and scales intact. Smell should be sea-fresh. **Fillets** Should be neat, firm, moist, trim, without discoloration or bruising. **Frozen** Should be hard with no sign of partial thawing or freezer-burn. Packaging should be intact.

Use fresh fish at once or store near bottom of refrigerator at 0-5°C for up to 2 days. To refrigerate, rinse, pat dry with kitchen paper, place on a plate and cover with clingwrap or another plate. Use by recommended date. **Canned fish** Transfer to another container after opening and store covered in refrigerator for up to 24 hrs.

Maximum storage period is 3 months. Freeze directly after purchase unless fish has previously been frozen: see label. To freeze, wash fish and pat dry with kitchen paper, double-wrap in polythene, freezer-wrap or foil for extra protection and pack in freezer bags. Expel as much air as possible and store at −18°C or below. Thaw overnight in refrigerator or in microwave using defrost setting. Do not thaw in water. Never refreeze fish after thawing.

| Preparing and cooking | Typical uses | Alternative forms |
|---|---|---|
| Gut and clean before cooking. Some fish should be scaled first. Skinning is optional. Cooking time depends on type and cooking method. Overcooking makes fish tough, dry and unpleasant. Bake, braise, deep-fry, grill, microwave, casserole, poach, shallow-fry, steam or stir-fry. Cooked fish looks opaque or white in colour and a fork or skewer easily pierces flesh. Overcooked fish is dry and falls apart easily. | Species of same type are usually interchangeable. Serve with a sauce, as fish pie, fishcakes, kebabs, mousses, pâtés, salads. | Dried, salted, 'exotic' fish (e.g. swordfish and shark), frozen fish products (e.g. fish fingers), deli products (e.g. pâté), chilled fish products, dehydrated fish meals, dips, paste, liver, oil, sausages, smoked. |
| As above. Cooked fish flakes easily when tested with a fork. | Species of same type are usually interchangeable. Serve as is or use in salads, pizzas, on toast, in oatmeal, sandwiches, flans etc. | Deli products (e.g. pâté), canned in water, brine, oil or sauce (e.g. tuna, sardines, salmon), sardines, anchovies, salmon, paste, dips, pickled, dried, salted, cured, sausages, dehydrated fish meals. |

| Food | How to choose | Storage | Freezing |
|------|---------------|---------|----------|
| **Smoked fish** Fish is either hot- or cold-smoked after immersion in brine solution. Hot-smoked fish (e.g. mackerel), smoked at high temperature, is cooked and ready to eat. Cold-smoked fish (e.g. kippers) has smoky flavour but is still raw so needs cooking. Cod, haddock, herring (kippers), mackerel, trout and salmon are often smoked. | Buy from refrigerated counters or cabinets. Check 'use by' date. Fish should have fresh, pleasant, smoky aroma and glossy appearance. Flesh should be firm, not sticky, ragged or discoloured. | Slightly longer shelf-life than fresh fish. Store away from other food as smell and dye can be absorbed. Keep well-wrapped above raw food in refrigerator at 0-5°C for up to 3 days. Use by recommended date. | As oily, oil-rich/fatty fish. |
| **Shellfish/seafood** 3 main groups: crustaceans, molluscs and cephalopods. Types include crab, lobster, prawns, scampi (crustaceans), clams, cockles, mussels, oyster (molluscs), octopus, squid (cephalopods). | Available fresh, ready-cooked or frozen. Buy from a reliable supplier and check 'use by' date. Shells should be undamaged and tightly closed without cracks; open shells that do not close when tapped sharply should be discarded. Cooked shellfish should be firm to touch. 'Live' seafood (e.g. lobsters and crabs) should look lively and be bright and moist, not pale and dry. They should have all their limbs, be a good colour and heavy for their size. | Cooked shellfish is best used on day of purchase as it deteriorates rapidly, otherwise refrigerate overnight above raw food in refrigerator at 0-5°C. **Live shellfish** Wash, dry and place in bowl or on plate, then cover with a damp cloth and another plate or clingwrap. Use by recommended date. | Maximum storage period is 3 months. Freeze directly after purchase unless seafood/shellfish has previously been frozen: see label. Do not freeze fresh shellfish unless it is very fresh. Live shellfish which has been cooked at home may be frozen. Ideally, before freezing, remove meat from shells. Pack seafood/shellfish in airtight freezer bags or containers. Expel as much air as possible and store at −18°C or below. Thaw overnight in refrigerator or in microwave using defrost setting. Do not thaw in water. Never refreeze after thawing. |

| Preparing and cooking | Typical uses | Alternative forms |
|---|---|---|
| Skinning is optional. Cooking time depends on type and cooking method. Overcooking makes fish tough, dry and unpleasant. Bake, braise, deep-fry, grill, microwave, casserole, poach, shallow-fry, steam or stir-fry. Cooked fish flakes easily when tested with a fork. Overcooked fish is dry and falls apart easily. | Species of same type are usually interchangeable. Serve as is or use in kedgeree, soups, sauces, sandwiches, pâtés, pastes. | Deli products (e.g. pâté), paste, dips. |
| Versatile serving methods include sautéing, steaming or boiling, deep-frying. Never overcook. Cooking times depend on type and cooking method. **Crustaceans** (e.g. crabs and lobsters) Boil for up to 35 mins and allow to cool, then prepare and use in salads and hot dishes. For crabs, allow 15 mins for first 450g/1lb and 10 mins for each 450g/1lb thereafter, up to a maximum of 35 minutes. **Cephalopods** Clean, cut into strips and either batter and fry quickly, or use in stir-fries. | Dressed crabs, kebabs, lobster thermidor, prawn cocktail, salads, soups, stir-fries. | Smoked scallops, smoked prawns, dried, frozen, salted, spice-cured, canned in brine or oil, used in dehydrated shellfish meals, dips, paste, pâté, sausages, spreads, or pickled in vinegar or brine. |

| Food | How to choose | Storage | Freezing |
|------|---------------|---------|----------|
| **FRUIT** **Citrus fruit** Examples (all imported): clementine, grapefruit, kumquat, lemon, lime, limequat, mandarin, minneola, orange, ortanique, pomelo, satsuma, shaddock, tangerine, ugli fruit. Some, such as grapefruit, are available in more than one variety including those with yellow or pink flesh. Some are seedless. All are rich in vitamin C. | Check 'display until' date (if shown). Choose fruits with bright, uniform colour that feel heavy for their size. Avoid those with shrivelled or bruised skin. Most varieties should have tight, slightly rough skins. Others have loose, easy-peel skins. | Generally, citrus fruits keep well. Keep at room temperature for up to 1 week or store in refrigerator at 0-5°C for up to 1 month. | Lemon, lime or orange rind can be frozen mixed with a little fruit juice or water in an ice-cube tray. Lemon and orange slices can be frozen and added straight to drinks when required. Peeled and segmented flesh may be frozen at −18°C or below in syrup or with sugar for up to 1 year. |
| **Dried fruit** Examples: apples, apricots, bananas, blueberries, currants, dates, kiwi, mango, peaches, pears, pineapple, prunes, raisins, sultanas. Available dried, semi-dried or ready-to-eat. Also sold in mixed packs. | Check 'best before' date. Choose intensely coloured fruit. Avoid dull or dusty dried fruit or damaged, split packets. Fruit is sometimes treated with a preservative such as sulphur dioxide to extend shelf-life of fruit: see label. | **Unopened** Keep in packet. **Opened** Store in airtight container in cool, dry, airy place. Most keep well for up to 18 months. Use by recommended date. | Not applicable. May be frozen in dishes. |

## Preparing and cooking

## Typical uses

## Alternative forms

Wash before use, if using the skin or rind. Many varieties can be peeled and eaten raw. Others, such as kumquats and limequats, can be eaten whole, skin and all. Lemons and limes add flavour and tang to dishes and can usually substitute for one another in dishes. Lemon juice is often used to prevent flesh of other fruit and vegetables from discolouring.

Eaten in casseroles, compotes, confectionery, dressings, drinks, eaten raw, used in flans and tarts, fruit salad, as garnish or decoration, grilled with honey, jellies, juice used for marinades and salad dressings, lemon meringue pie, marmalades and preserves, mousses, sorbets; grate rind to add flavour to baked goods such as biscuits, cakes and puddings.

Candied and dried peel, candied fruit, canned in fruit juice or syrup, essence or oil, frozen fruit juice concentrate, fruit juice, lemon, lime or orange curd, lemon slices in jars, marmalades and preserves, organic, pickled or preserved.

Ready-to-eat can be used as they are. Others need soaking first. Cook as part of dish or poach before serving.

In baked goods such as biscuits, breads, cakes, desserts and puddings, casseroles, Christmas cake, Christmas pudding, chutneys, compotes, currant/fruit buns, fools, ice-cream, mincemeat, mousses, muesli, purées, sauces, spotted dick, stews, stuffing, teabreads.

| Food | How to choose | Storage | Freezing |
|------|---------------|---------|----------|
| **Orchard fruits** Fruits grown on trees in orchards which contain small seeds or a stone at their core. Examples: apple, apricot, cherry, damson, greengage, nectarine, peach, pear, plum, quince. | Check 'display until' date (if shown). When ripe, softer fruits deepen in colour and should be soft to touch at stalk end. **Apples** Should be firm, crisp, bright in colour. Avoid wrinkled, bruised skin with soft patches. **Apricots** Fruit should be firm with unwrinkled, velvet-soft skin and warm deep colour. **Cherries** Vary in colour from white to deep purple. Fruit should be plump, shiny, firm and colourful, not soft, bruised or split. Stems of fresh cherries are green and flexible: avoid those with dry, brittle brown stems. Ideally, eat on day of purchase. **Damsons** Should have smooth skin, uniform colour, slightly soft flesh. **Greengages** As damsons. **Nectarines and peaches** Choose firm, not hard, ones with yellow or cream background colour. Avoid green, bruised or wrinkled skins and brown patches. White peaches have best flavour. **Pears** Should be firm, slightly softer at stalk end, stalk intact. **Plums** Many varieties, shapes and colours. Choose ones with smooth skin, uniform colour and slightly soft flesh. **Quince** Golden in colour when ripe. Choose plump, hard fruit with unblemished skin and no soft patches. | Softer fruits are more perishable. Handle carefully as they bruise easily. Some may be stored in refrigerator at 0-5°C. **Apples** Store wrapped in paper for several months in dark, cool, airy place (crispness and flavour decline after some time). Refrigerate for up to 3 weeks. **Apricots** Ripen at room temperature. Refrigerate for up to 1 week. **Cherries** Ripen at room temperature. Perishable, best eaten on day of purchase. Refrigerate. **Damsons** Ripen at room temperature. Place in open plastic or paper bag and refrigerate for up to 1 week. **Greengages** As damsons. **Nectarines and peaches** As damsons. **Pears** Ripen at room temperature. Place in open plastic or paper bag and refrigerate for up to 2 weeks. **Plums** As damsons. **Quinces** As pears. | Store in freezer at −18°C or below. **Apples** Peel and blanch for 2-3 mins. Cool and pack with or without sugar. Freeze for up to 1 year. **Apricots** Freeze raw or cooked in syrup for up to 1 year. **Cherries** Prepare and stone first. Freeze raw or cooked for up to 1 year, on own or with sugar or syrup. **Damsons** Freeze raw or cooked for up to 1 year, on own or in syrup. **Greengages** Halve and stone fruit. Freeze raw or cooked for up to 1 year. **Nectarines and peaches** Peel and freeze in syrup for up to 1 year. **Pears** Peel and freeze in syrup for up to 1 year. Very soft when thawed. **Plums** Halve and stone fruit. Freeze raw or cooked for up to 1 year. **Quinces** Peel and blanch for 2-3 mins. Cool, pack and freeze for up to 1 year. |

## Preparing and cooking

Wash before use or eating raw.
**Apples** Dessert varieties can be eaten raw or cooked (baked, grilled, poached, sautéed). Tarter cooking apples are usually cooked and often sweetened. Crab-apples are suitable only for cooked dishes and preserves. All discolour when cut so coat with lemon juice or store in acidulated water. Apples have an affinity with cinnamon and cloves.
**Apricots** Eat raw or baked, grilled, lightly poached. **Cherries** Eat raw or lightly cooked (baked, poached, sautéed, stewed). **Damsons** Best served stewed or lightly cooked.
**Greengages** Eat raw or cooked (baked, grilled, poached).
**Nectarines and peaches** Eat raw or lightly cooked (baked, grilled, poached). Flesh discolours when cut. **Pears** Eat raw or baked, grilled, gently poached, sautéed. Discolour when cut so coat with lemon juice or store in acidulated water. **Plums** As greengages. **Quinces** Peel, core and seed before cooking.

## Typical uses

Baked, batter puddings, breads, cakes, casseroles, chutney, compotes, desserts and puddings, flans, fruit salads, gâteaux, ice-creams, jams, jellies, paste, pies, poached, preserves, purées, salads, sauces, sorbets, soups, stews, tarte Tatin, tarts. Quinces are used mainly for jams and preserves, and in some savoury dishes.

## Alternative forms

Apple butter, apple sauce or jelly, bottled juice, candied, canned in fruit juice or syrup, cherry liqueur, chutney, damson cheese, dried or semi-dried (apples, apricots, cherries, peaches, pears, plums), frozen, glacé cherries, in jars, jam, maraschino cherries, organic, pickle, pickled, pie filling, preserves, purées (e.g. apple purée).

| Food | How to choose | Storage | Freezing |
|---|---|---|---|
| **Soft fruit**<br>Mainly berry fruits. Examples: bilberry, blackberry, blueberry, boysenberry, cranberry, currant (black, red, white), elderberry, gooseberry, loganberry, mulberry, raspberry, rhubarb, strawberry, tayberry. | Check 'display until' date (if shown). **Berries** Should be plump and shiny, have intact hulls or stalks, a uniform colour and no signs of mould or rot. **Currants** As above. Avoid wrinkled or very soft fruit. **Rhubarb** Choose crisp, firm, deep-coloured stalks, not too large. | Most are perishable and should be eaten as soon as possible after purchase. Rhubarb and cranberries tend to keep for longer. Berry fruits are particularly prone to mould: pick out and discard damaged or mouldy berries. Handle fruit gently. Top and tail, hull or remove stalks before use. Store in refrigerator at 0-5°C. **Cranberries** Store in a punnet or open plastic or paper bag in the refrigerator for up to 2 weeks. **Rhubarb** Store in open plastic or paper bag for up to 5 days in refrigerator. **Strawberries** Keep covered in the refrigerator as their smell may taint and penetrate other foods. **Other berries and currants** Wrap loosely in open plastic or paper bag and keep in refrigerator for 3-5 days. | Store in freezer at −18°C or below. **Berries** Open-freeze, when firm pack in sealed containers or freezer bags. Freeze raw or cooked for up to 1 year. **Currants** As above. **Rhubarb** Trim, slice, pack and freeze for up to 1 year. Can be frozen in syrup. May lose some of its colour. |

## Preparing and cooking

Wash gently before use or eating raw. Many (e.g. blackberries, loganberries, raspberries and strawberries) can be eaten raw; others such as currants, rhubarb and gooseberries are better lightly stewed or cooked and sweetened before eating. Never eat rhubarb raw: always bake or stew before cooking. Rhubarb leaves are poisonous.

## Typical uses

Blueberry muffins, chilled or hot soups, cobblers, compotes, cranberry sauce, crumbles, decoration, fools, fruit salads, fruit tarts and flans, ice-creams, jams, jellies, mousses, pavlova, pies, preserves, purées, sauces, sorbets, summer pudding.

## Alternative forms

Canned or bottled in syrup or fruit juice, cassis liqueur, dried or semi-dried, frozen, as fruit juice, in jars, organic, pie-filling, relish.

| Food | How to choose | Storage | Freezing |
|---|---|---|---|
| **Tropical and exotic fruit**<br>Exotic fruit imported from tropical climates can now be found on market stalls and in supermarkets. Examples: banana, Cape gooseberry (physalis), custard apple (cherimoya), date, fig, guava, lychee, mango, mangosteen, papaya (paw-paw), persimmon (sharon fruit), pineapple, pomegranate, prickly pear, rambutan, star fruit (carambola), tamarillo. | Check 'display until' date (if shown). **Bananas** Choose plump fruits with unmarked green to yellow skins, no black patches. **Cape gooseberries** These smooth, firm, shiny berries are bright orange when ripe. Choose ones with undamaged husks. **Custard apples** Look for firm, unblemished fruits. **Dates** Should be plump and shiny with golden-brown skins, have firm, not dry flesh and smell fresh. **Figs** Vary in colour from green to purple or black. When ripe, figs are soft to touch and have a bloom on the skin. Over-ripe figs shrink. **Guavas** Look for firm, unblemished fruit, yellow when ripe with no dark blemishes. **Lychees** Should have rough, pinkish brittle shell with soft, white fragrant flesh. Under-ripe fruit is pale-beige or greenish, over-ripe is brown. Avoid fruit with dry, shrivelled or mouldy skin. **Mangoes** Many varieties, sizes and colours. Choose sweet-smelling fruits with tight, smooth skin free of soft patches. **Mangosteens** Choose purply-brown fruits with deep, bright, firm skin with no blemishes. **Papayas** Look for deep-coloured small ones. **Persimmons** Many varieties. Bright orange when ripe, should have stalk attached, soft and thin skin. **Pineapples** Vary in size and colour. Choose large, heavy fruit with sweet aroma and good colour with no soft patches. **Pomegranates** Choose hard fruits with unblemished skins, should feel heavy in weight. **Prickly pears** Fruits of the cactus. Deep orange when ripe. **Rambutans** Choose fruit with fresh-looking spines. Avoid fruit with matted, flat or brown spines. **Star fruit** Choose firm, unblemished fruit. **Tamarillos** Should have firm, smooth, satiny, blood-red or bright yellow skin and intact stalk. | Some can be refrigerated at 0-5°C. **Bananas** Store at room temperature. Do not refrigerate. **Cape gooseberries** Refrigerate for up to 3 days. **Custard apples** Refrigerate for up to 4 days. **Dates** Refrigerate for up to 2 weeks. **Figs** Do not keep well. Eat as soon as possible after purchase. Refrigerate for up to 2 days. **Guavas** Eat as soon as possible after purchase. **Lychees** Store at room temperature for up to 4 days. **Mangoes** Refrigerate for up to 2 weeks or at room temperature for up to 3 days. **Mangosteens** Refrigerate for up to 3 days. **Papayas** As mangoes. **Persimmons** Refrigerate for up to 2 days. **Pineapples** Ripen at room temperature. Store in plastic bags to prevent smell permeating other food and refrigerate for up to 3 days. **Pomegranates** Whole: refrigerate for up to 3 weeks. Kernels: refrigerate for up to 1 week. **Prickly pears** As mangosteens. **Rambutans** As lychees. **Star fruit** As mangosteens. **Tamarillos** As mangosteens. | Store in freezer at −18°C or below. **Bananas** Not suitable for freezing. **Cape gooseberries** As bananas. **Custard apples** As bananas. **Dates** Dry-pack and freeze for up to 1 year. **Figs** Freeze whole or peeled in syrup, for up to 1 year. **Guavas** Peel and freeze raw, cooked or puréed, or in syrup or sugar for up to 1 year. **Lychees** As guavas. **Mangoes** As guavas. **Mangosteens** As bananas. **Papayas** As guavas. **Persimmons** Peel and dry-pack or freeze with sugar or puréed. Freeze for up to 1 year. **Pineapples** Peel and freeze raw on own or in syrup or sugar for up to 1 year. **Pomegranates** Freeze kernels for up to 1 year. **Prickly pears** As bananas. **Rambutans** As guavas. **Star fruit** As bananas. **Tamarillos** As bananas. |

## Preparing and cooking

Wash before use (where applicable). **Bananas** Eat raw or lightly cooked, puréed, sautéed, baked or barbecued. Use at once after peeling or toss in lemon juice to prevent discoloration. **Cape gooseberries** Eat raw, use as a garnish. **Custard apples** Best eaten raw. Use in sorbets, fools, ice-cream, desserts. **Dates** Eat raw or stuffed. Use in breads, cakes, teabreads, compotes, salads. **Figs** Skin, flesh and seeds are all edible. Serve at room temperature. May be lightly poached or baked. **Guavas** Eat raw or cooked, puréed for sauces or added to savoury dishes. Use in ice-cream, sorbets, jams, jellies. **Lychees** Peel and eat raw or lightly poached. **Mangoes** Peel and eat raw, cooked or puréed. **Mangosteens** Eat flesh raw with a spoon. **Papayas** Skin, deseed and eat raw. Pairs well with coconut and lime. **Persimmons** Eat when really soft and ripe, otherwise they will be bitter. Eat raw or cooked, puréed for mousses and fools. **Pineapples** Remove peel, eyes and woody stem before eating. Eat raw or lightly cooked, grilled or sautéed. Enzymes in fruit prevent gelatine setting. **Pomegranates** Scoop out seeds and juice and eat raw or add to fruit salads. **Prickly pears** Covered in prickly hairs, so wear rubber gloves when handling them. Peel, slice and serve raw in fruit salads or compotes, or serve stewed and puréed. **Rambutans** Peel and eat raw or lightly poached. **Star fruit** Slice and eat raw. Poach lightly or use as decoration. **Tamarillos** Serve peeled and lightly stewed with sugar. Eat raw. Use in ice-cream, sauces, pickled.

## Typical uses

**Bananas** Baked, breads, cakes, deep-fried, drinks, flambéed, ice-cream. **Cape gooseberries** Jams, puréed; as decoration. **Custard apples** Puréed to make ice-cream. **Dates** Biscuits, breads, chutneys, fruit salads, stuffed. **Figs** Fruit salads, fruit tarts, jams, soufflé. **Guavas** Ice-cream, jams, jellies, salads, sorbets. **Lychees** Fruit salads, served with cream or ice-cream. **Mangoes** Chutneys, desserts, puréed, salads, sauces, served with smoked meats or cooked poultry. **Mangosteens** Eaten plain, sometimes pickled. **Papayas** Chutneys, salads, stuffed. **Persimmons** Accompaniment to cheese or meat, custards, mousses, topping ice-cream. **Pineapples** Fruit salads, grilled, kebabs, upside-down cake. **Pomegranates** Fruit salads, ice-cream, mousses, sauces. **Prickly pears** Puréed for custards, ice-cream, jams, pastes. **Rambutans** Fruit salads, served with cream or ice-cream. **Star fruit** Fruit salads, garnish, served with poultry or shellfish. **Tamarillos** Jams, ice-cream, pickled, raw or baked, sauces.

## Alternative forms

Bottled, canned in fruit juice or syrup, dried or semi-dried, fruit juice, marinated figs, nectar, paste, pickles and chutneys, pie fillings, preserves, sauces, sugared. **Banana** Dried chips. **Dates** Boxed or dried in block, dried, sugared and chopped.

| Food | How to choose | Storage | Freezing |
|------|---------------|---------|----------|
| **Vine fruit** Fruits of climbing, twining plants or vines. Examples: grapes, grenadilla, kiwi, melon, passionfruit. | Check 'display until' date (if shown). **Grapes** Vary in colour and size (yellow, green, red, black). Some are seedless. Choose plump, firm fruit, well attached to stalk. Avoid bruised, blemished or mouldy fruit. **Grenadillas** Look for firm, smooth, unblemished skin and stiff stalks. **Kiwi fruit** Choose firm fruits, slightly soft when pressed. Avoid fruit with soft patches. **Melons** Many varieties, most smell fragrant when ripe. The more mature, the sweeter. Should feel heavy for their size. Choose firm fruit without soft patches or cracks. **Passionfruit** Should look wrinkled and purple if ripe, and be heavy for their size. | Some can be refrigerated at 0-5°C. **Grapes** Refrigerate in open plastic or paper bag for up to 2 weeks. Store at room temperature for a few days. **Grenadillas** Refrigerate for up to 3 days. **Kiwi fruit** Refrigerate for up to 2 weeks. **Melons** Store at room temperature for up to 3 days. After slicing, remaining seeds should be left in melon to impede dehydration and oxidation. Cover tightly to prevent flavour penetrating other foods, and refrigerate for up to 3 days. **Passionfruit** As grenadillas. | Store in freezer at −18°C or below. **Grapes** If seedless, freeze whole. Otherwise, remove seeds and freeze. May also be frozen in syrup. **Grenadillas** Peel and dry-pack with sugar or syrup. Freeze for up to 1 year. **Kiwi fruit** As grenadillas. **Melons** Not generally suitable for freezing. Some types, e.g. cantaloupe and honeydew, freeze quite well in syrup, or dry-packed. They lose crispness on thawing. **Passionfruit** As grenadillas. |

## GRAINS, CEREALS AND FLOURS

| | | | |
|------|---------------|---------|----------|
| **Grains and cereals** Edible grains from food plants include barley, bran, buckwheat, bulgar wheat, corn, cornmeal, couscous, cracked wheat, farina, grits, groats, hominy, oats, polenta, rye, sago, semolina, tapioca, wheatgerm. | Check 'best before' date. Buy as fresh as possible from a reputable source with a high turnover. Fresh grain is plump and dry with bright, even colouring. Avoid crumbly or shrivelled grain with a musty or sour odour. | Store in airtight container in cool, dry place for up to 1 year. Whole grain may become infested with weevils, which are undesirable but totally harmless. Use by recommended date. | Can be frozen in dishes. |

| Preparing and cooking | Typical uses | Alternative forms |
|---|---|---|
| Wash before use (as appropriate). **Grapes** Eat raw or serve hot, lightly poached. **Grenadillas** Halve and scoop out, eat flesh and seeds. **Kiwi fruit** Cut in half or cut the top off and eat with a teaspoon. Peel and slice or chop. Enzymes in fruit prevent gelatine setting. **Melons** Serve smaller melons in halves, larger melons in wedges with seeds scooped out. **Passionfruit** As for grenadillas. | **Grapes** Accompany fish and light meats such as chicken, used in fruit tarts, fruit salads, jellies, open tarts, served with cheese. **Grenadillas** Desserts, sauces, sprinkled over salad. **Kiwi fruit** Fruit salads, fruit tarts, pavlova. **Melons** Desserts, fruit salads, served with Parma ham, soups, sprinkled with port or ginger. **Passionfruit** As for grenadillas. | Bottled in syrup, dried (mainly grapes: raisins, currants, sultanas), dried or semi-dried kiwi fruit, frozen, grape juice, jams, jellies, pickled or candied rind (melon), preserves, wine. |
| Before cooking, rinse in cold water to remove dust or dirt particles. Boil, bake or steam. | Biscuits, bread, breakfast cereals, casseroles, crumble toppings, dumplings, herrings in oatmeal, milk, moulds, oatcakes, pancakes, popcorn, porridge, puddings, semolina pudding, snacks, soups, stuffing, tabbouleh. | Barley flour, barley water, canned or frozen hominy, corn chips, corn oil, corn syrup, cornflour, instant oatmeal, instant polenta, malted barley, oat flakes, oatcakes, oat milk, organic flakes and oats, pearl barley, puffed oats, quinoa, rye flakes, semolina flour, triticale, wheat flakes. |

| Food | How to choose | Storage | Freezing |
|------|---------------|---------|----------|
| **Rice**<br>Two main groups: all-purpose and speciality. All-purpose includes long-grain, white or brown and easy-cook long-grain. Speciality varieties include arborio, basmati, glutinous, jasmine, pudding (short-grain), wild. | Check 'best before' date. Buy appropriate type. Good-quality brown or white long-grain rice is useful store-cupboard item. | Rice keeps for up to 1 year. Brown rice keeps for less time than white as grains contain natural oil, which may turn rancid. Use by recommended date. **Unopened** Store in packet in cool, dry cupboard. **Opened** Store in airtight container as above. **Cooked rice** Cool quickly, place in covered or airtight container and refrigerate for up to 24 hrs. | Store in freezer at −18°C or below depending on suitability. **Raw rice** Not applicable. **Cooked rice** Cool quickly, place in freezer bags. Open-freeze and store in sealed bags or containers; the frozen rice will pour freely as required. **Plain cooked rice** Seal and freeze for up to 3 months. **Rice dishes** (such as risotto) Freeze for up to 6 weeks. |
| **Flours**<br>Main types are derived from grains e.g. barley, buckwheat, millet, rice, rye and wheat flours, sold as plain, self-raising, sponge, strong, white, wholemeal, malted brown. Other types are made from seeds, roots or tubers. Flours vary in taste, nutritional value and function. | Check 'best before' date. Avoid opened or damaged packages. Choose appropriate type of flour for purpose, e.g. self-raising flour for cakes, strong flour for bread. | Store unopened in a cool, dark, dry place. Store opened in airtight container as above for up to 6 months. Use by recommended date. | May be frozen in sealed plastic freezer bags or plastic airtight containers for several months. Store in freezer at −18°C or below. Thaw at room temperature. |

## Preparing and cooking

Some rice needs washing before being cooked. See label. Allow at least 55g (2oz) rice per person. Rice trebles in size as it cooks so cook in plenty of boiling water. Simmer on hob or cook in oven. Drain off excess water at end of cooking time. Can also be cooked in microwave oven. Always reheat cooked rice quickly and thoroughly until piping hot. Do not reheat cooked rice more than once. Soaking rice reduces its cooking time. Cook frozen rice in boiling water, in the oven or microwave. Use to accompany meals or in dishes. **All-purpose rice** Use in Chinese, Mexican, Spanish and American cooking, in stir-fries, rice salads, stuffings, paella, pilaff, fried rice, biryani, kedgeree, jambalaya. **Speciality rice** Use for risotto, rice puddings, sushi, rice puddings, rice salads, stir-fries.

Always cooked with other ingredients or incorporated into dishes. Wheat flour is used particularly in baked dishes and to thicken sauces. White and wholemeal flours are usually interchangeable in recipes as long as appropriate type (e.g. self-raising for sponge cakes, strong for yeast cookery) is used.

## Typical uses

To accompany main courses such as chilli con carne or curry, kedgeree, paella, rice salads, risotto, stir-fries, stuffings; sweet dishes include rice puddings and creamed rice.

Batters, biscuits, bread, cakes, crêpes, crumble toppings, flatbreads, pancakes, pastries, pizza bases, sauces, scones, teabreads.

## Alternative forms

Boil-in-the-bag, canned, frozen, ground, puffed, flaked, flavoured rice mixes, non-dairy beverages made from brown rice, pilau, quick-cook, rice bran, rice cakes, rice crackers, rice flour, rice noodles, rice paper, rice pasta, rice pudding; rice beer, sake, vinegar, wine.

Arrowroot, batter mix, bread mix, buckwheat flour, cake and biscuit mixes, carob flour, chickpea flour, cornflour, kelp flour, millet flour, organic flour, pastry and dumpling packet mixes, potato flour, semolina flour, soya flour (gluten-free), spelt flour.

| Food | How to choose | Storage | Freezing |
|------|---------------|---------|----------|

## MEAT AND POULTRY: MEAT

**Bacon, gammon and ham**

Bacon, gammon and ham are pork flesh which has been cured by salting. Available smoked and unsmoked. (Unsmoked bacon is also known as green bacon.) Sold vacuum-packed or loose, in a variety of cuts, cooked and uncooked, including joints, rashers, slices, steaks.

Buy from a reputable source, from refrigerated cabinet. Check 'use by' date. Fat should look firm and pale-white or cream in colour. **Gammon and bacon** Should be slightly moist, clear red-pink and even-coloured throughout. **Bacon rashers** Sold with or without rind. **Ham** Varies in colour from pale to dark pink depending on type; should be moist but firm, depending on cure. Check label and curing method used, e.g. tender-sweet.

Refrigerate directly after purchase at 0-5°C. Remove meat bought loose from packaging and seal in a clean polythene bag. Treat opened vacuum packs as fresh meat. Use by recommended date. **Gammon and bacon** Refrigerate uncooked rashers or joints for up to 1 week. **Vacuum-packed bacon** Refrigerate for 2-3 weeks. **Ham** Refrigerate cooked or sliced ham for up to 2 days.

Freeze directly after purchase at −18°C or below. Storage periods vary: due to high salt content bacon and ham do not keep as long as fresh meat. Dishes containing bacon or ham can be frozen for up to 1 month. Vacuum-packed meat keeps longest. **Uncooked meat** Freeze for up to 1 month. **Vacuum-packed uncooked meat** Freeze for up to 3 months. **Gammon and bacon** Freeze joints for up to 3 months. **Ham** Cooked ham is not ideal for freezing but can be frozen for up to 1 month. Freeze vacuum-packed ham for up to 3 months.

**Beef**

Meat of mature cattle. Cuts include boneless rolled joint, braising steak, brisket, cubed, fillet steak, flash-fry or quick-fry cuts, fore-rib or rib roast, minced or ground beef, rump steak, silverside, sirloin joint and steak, stewing steak (shin, leg, neck), stir-fry, thick flank, topside.

Buy from a reputable source, from refrigerated cabinet. Check 'use by' date. Well-hung beef is deep red in colour, tender and flavoursome. Choose lean meat trimmed of excess fat. The remaining fat should be creamy-white in colour. Avoid meat with an unpleasant smell, yellow fat, a shiny surface, greenish tinge or dry edges.

Refrigerate directly after purchase at 0-5°C. Refrigerate pre-packed meat in original packaging, if the packaging is clean and dry. If not or if bought from a butcher, unwrap, place in a covered container or dish covered tightly with clingwrap or foil and refrigerate. Do not allow meat juices to drip from packets or dishes. Wrap cooked meat in clingwrap or foil. Cover and refrigerate uncooked meat separately and lower in the refrigerator than cooked meat to avoid cross-contamination. Chill leftover meat or meat dishes as quickly as possible. Use by recommended date. **Cooked/uncooked joints** Refrigerate for up to 3 days. **Uncooked mince** Refrigerate for up to 1 day. **Sliced meat, pâté, casseroles and stock** Refrigerate for up to 2 days. **Meat pies** Refrigerate for up to 1 day.

Freeze directly after purchase at −18°C or below. Wipe fresh meat with absorbent kitchen paper before packing and freezing, to remove excess moisture. Cover with foil or clingwrap before freezing or seal in a freezer bag or airtight container. Wrap individual portions separately if desired. Defrost frozen meat thoroughly. Do not refreeze after thawing. **Whole joints or pieces** Freeze for up to 9-12 months. **Mince** Freeze for up to 3 months. **Cooked meat dishes** Freeze in airtight container for up to 3 months.

## Preparing and cooking | Typical uses | Alternative forms

Baking, boiling, braising, frying, grilling, roasting, simmering, stewing. Some bacon and gammon joints may need soaking before cooking, depending on curing method and salt content. See label.

Burgers, kebabs, quiche lorraine, soups, stews. Bacon is often used for rolling or wrapping round foods, e.g. devils on horseback.

Canned, bacon bread, meat paste, raw air-cured ham such as parma and prosciutto, smoked.

Fresh meat should not need to be washed before cooking. Use absorbent kitchen paper to dry off excess moisture before frying, stir-frying or roasting. This will reduce spitting. Cooking methods include barbecuing, boiling, braising, casseroling, frying, grilling, kebabs, microwaving, pot-roasting, roasting, stir-frying. Use meat thermometer when roasting. Reheat cooked meat dishes once only. Reheat thoroughly until piping hot. Use sharp knife for cutting or carving.

Boeuf bourguignon, beef olives, beef Wellington, casseroles, chilli con carne, Cornish pasties, curries, filling for tacos, hamburgers or beef burgers, lasagne, meatloaf, meatballs, pies, roast joint, shepherd's pie, spaghetti bolognese, steak and kidney pie, stews, stir-fries, stroganov, stuffed vegetables.

Beef products and meals, canned, corned, dehydrated ready-meals, dried, extra lean or prime cuts, fresh and frozen beef burgers, meat paste, organic, pastrami, pâtés, peppered, pickled, potted, ready-marinated, ready-prepared kebabs, beef olives, reduced-fat beef burgers and sausages, salted, sausages, spiced.

| Food | How to choose | Storage | Freezing |
|------|---------------|---------|----------|
| **Lamb**<br>Meat from new or old season's lamb. Cuts include best-end, boneless rolled joint, breast, chump and loin chops, crown roast, cubed, cutlets, guard of honour, leg joint (gigot), leg steaks, middle neck, minced/ground lamb, neck fillet, noisettes, riblets, shoulder. | Buy from a reputable source, from refrigerated cabinet. Check 'use by' date. Look for fine-grained lean meat, pinkish-brown in colour, with fat that is firm, dry and creamy in colour. Avoid meat with an unpleasant smell, yellow fat, a shiny surface, greenish tinge or dry edges. New season's lamb is most succulent, particularly that of 3-9 months old. | As beef. | Freeze directly after purchase at −18°C or below. Wipe fresh meat with absorbent kitchen paper before packing and freezing, to remove excess moisture. Cover with foil or clingwrap before freezing or seal in a freezer bag or airtight container. Wrap individual portions separately if desired. Defrost frozen meat thoroughly. Do not refreeze after thawing. **Whole joints or pieces** Freeze for up to 6-9 months. **Mince** Freeze for up to 3 months. **Cooked meat dishes** Freeze in airtight container for up to 3 months. |
| **Offal**<br>Edible internal organs and parts of animal remaining after main meat sources have been removed. Examples: cheeks, ears, feet, head, heart, kidney, liver, stomach, tail, tongue, tripe. Highly perishable. | Buy from a reputable source, from refrigerated cabinet. Check 'use by' date. Freshness is crucial. Offal should look and smell fresh (not strong), be moist and shiny with no dry patches. Avoid offal with greenish colour or dry surface. | Refrigerate directly after purchase at 0-5°C. Store uncooked for up to 1 day. Ideally use on day of purchase. Use by recommended date. | Freeze directly after purchase at −18°C or below. Remove all fat, skin and tubes, then wash and dry thoroughly before packing and freezing. Cover with foil or clingwrap before freezing or seal in a freezer bag or airtight container. Wrap individual portions separately if desired. Freeze for up to 3 months. Defrost meat thoroughly. Do not refreeze after thawing. |

| Preparing and cooking | Typical uses | Alternative forms |
|---|---|---|
| Fresh meat should not need to be washed before cooking. Use absorbent kitchen paper to dry off excess moisture before frying, stir-frying or roasting. This will reduce spitting. Cooking methods include barbecuing, braising, casseroling, frying, grilling, microwaving, pot roasting, roasting, stir-frying. | Burgers, casseroles, cottage pie, curries, Irish stew, koftas, Lancashire hotpot, marinated, meatloaf, moussaka, pies, roast joints, shish kebabs, soups, stuffed vegetables. | Extra lean and prime cuts, organic, reduced-fat burgers and sausages, sausages, smoked, trimmed cuts. |
| Offal such as tripe must be blanched before cooking to clean it and strengthen texture. Some offal such as tongue needs to be soaked before cooking. Cooking methods include braising, deep-frying, frying, grilling, poaching, sautéing. | Brawn, devilled kidneys, pâté, savoury pies, steak and kidney pie, terrines, tripe and onions. | Ready-frozen, smoked or pickled, lights, marrowbone, mesentery, pig's caul, testicles. |

| Food | How to choose | Storage | Freezing |
|------|---------------|---------|----------|
| **Pork**<br>Flesh of pig. Cuts include belly slices, boneless rolled joints, chump chops and steaks, cubed or diced, hand, leg joint (gigot), leg and shoulder steaks, loin chops and steaks, minced/ground pork, spare rib joint and chops, spare ribs, stir-fry, tenderloin or fillet. | Buy from a reputable source, from refrigerated cabinet. Check 'use by' date. Meat should be firm, pale pink in colour and moist. When freshly cut the fat should be firm, white and thin. Avoid grey or red-coloured meat and meat with discoloured patches. | As beef. | Freeze as soon as possible after purchase at −18°C or below. Wipe fresh meat with absorbent kitchen paper before packing and freezing, to remove excess moisture. Cover with foil or clingwrap before freezing or seal in a freezer bag or airtight container. Wrap individual portions separately if desired. Defrost frozen meat thoroughly. Do not refreeze after thawing. **Whole joints or pieces** Freeze for up to 4-6 months. **Mince** Freeze for up to 3 months. **Cooked meat dishes** Freeze in airtight container for up to 3 months. |
| **Veal**<br>Meat from calves, either milk-fed or grass-fed. Flesh of milk-fed calves, very pale with a mild flavour, is more expensive. Flesh of grass-fed calves is cheaper, stronger in flavour and slightly tougher. Cuts include breast, escalope, knuckle, leg, loin chop, neck, rib chop, rump, scrag, shin, shoulder, top rump. | Buy from a reputable source, from refrigerated cabinet. Check 'use by' date. Meat should be pale pink, moist and finely grained. Fat, if present, should be creamy-white colour. Bones should be white or pink. Avoid reddish, brown or grey meat, and meat with dry or mottled appearance. | As beef. | Freeze directly after purchase at −18°C or below. Wipe fresh meat with absorbent kitchen paper before packing and freezing, to remove excess moisture. Cover with foil or clingwrap before freezing or seal in a freezer bag or airtight container. Wrap individual portions separately if desired. Defrost frozen meat thoroughly. Do not refreeze after thawing. **Whole joints or pieces** Freeze for up to 6 months. **Mince** Freeze for up to 3 months. **Cooked meat dishes** Freeze in airtight container for up to 3 months. |

| Preparing and cooking | Typical uses | Alternative forms |
|---|---|---|
| Fresh meat should not need to be washed before cooking. Use absorbent kitchen paper to dry off excess moisture before frying, stir-frying or roasting. This will reduce spitting. Cooking methods include baking, barbecuing, braising, frying, grilling, microwaving, roasting, sautéing, stir-frying. Cooking times depend on cut of meat and recipe, but pork should be medium to well done (with an internal temperature of 75-80°C/167-76°F for medium, 80-85°C/176-85°F for well done). | Burgers, casseroles, curries, goulash, kebabs, marinated, meatloaf, pies, rissoles, roast joints, satay, stir-fry, stuffed vegetables. | Canned, extra lean and prime cuts, extra lean mince, meat paste, organic, pâté, pickled, pork pies, reduced-fat burgers and sausages, salted, sausages, smoked. |
| Fresh meat should not need to be washed before cooking. Use absorbent kitchen paper to dry off excess moisture before frying, stir-frying or roasting. This will reduce spitting. Cooking methods include baking, boiling, braising, casseroling, frying, grilling, pot roasting, roasting, stewing. Bard roasts with bacon or pork fat to keep them moist. Roast or grill at lower temperatures than other meats. Veal has little natural fat. Veal and chicken recipes are often interchangeable. | Blanquette de veau, casseroles, osso bucco, pies, roasts, stews, stuffed, vitello tonnato, wiener schnitzel. | Sausages. |

| Food | How to choose | Storage | Freezing |
|------|---------------|---------|----------|

**MEAT AND POULTRY: MEAT PRODUCTS**

**Cooked sausages and cold meats**
Cured, air-dried or smoked. Numerous types and regional varieties include black pudding, bratwurst, cervelat, chorizo, corned beef, frankfurter, kabanos, liver sausage, liverwurst, mortadella, pastrami, peperoni, salami, tongue. Flavours vary but are usually distinctive; some are very spicy. Sold ready-to-eat, often pre-packaged.

Check 'use by' date. Whole cooked sausages should be uncut or cut to order and should be moist and smell fresh. Dried meats should be firm but not over-dry.

Refrigerate at 0-5°C for up to 2 weeks. Sliced sausages and meat: refrigerate for up to 3 days. Use by recommended date.

Not applicable.

**Sausages (raw)**
A sausage is a meat-based mixture stuffed into a casing. Varieties include beef, black pudding, chipolatas, cocktail sausages, lamb, pork and beef, pork, seasoned in different ways. Regional varieties include Cotswold, Cumberland, Lincolnshire, Oxford.

Check 'use by' date. Should be moist, pink and fresh-smelling, not grey or dry-edged. If bought ready-frozen, store and cook according to label.

Refrigerate uncooked sausages at 0-5°C for up to 3 days. Use within 2 days of purchase. If unopened, refrigerate in original packaging. Once opened, store in refrigerator in container covered with clingwrap or foil. Will turn rancid if stored too long or at too warm a temperature. Use by recommended date.

Freeze directly after purchase at –18°C or below. Freeze in original packaging for up to 3 months. Pork and flavoured pork sausagemeat are available ready-frozen.

| Preparing and cooking | Typical uses | Alternative forms |
|---|---|---|
| Barbecued, fried, grilled, pan-fried, poached, sautéed, stewed. Also eaten cold, sliced. Some are spreadable. | Slice and eat cold, or use in salads, Italian dishes such as pizza, stews. | Canned. |
| Cook thoroughly before eating. Cooking times vary according to size/thickness of sausage. Do not prick sausages before cooking. Grill under medium heat for 15-20 mins, turning regularly, until pale-brown all over and cooked through. Can also be baked, fried or deep-fried. | Battered, casserole, salads, sausage plait, sausage rolls, scotch eggs, toad-in-the-hole. | Cocktail sausages, fresh and frozen sausage meat, fresh free-range sausages, haggis, ready-prepared/cooked products such as sausage rolls, reduced- or low-fat, skinless. |

| Food | How to choose | Storage | Freezing |
|------|---------------|---------|----------|

**MEAT AND POULTRY: POULTRY**

**Chicken**

Types available include boiling fowl, broiler, capon, corn-fed, free-range, organic, poussin (see page 98); fresh or frozen, cooked or uncooked. May be sold whole (with or without giblets) or in portions such as breast fillets, drumsticks, half chicken, leg and breast joints, quarter joints, thighs, wings. Some chicken is sold oven-ready, i.e. plucked and drawn. Wide range of ready-prepared chicken products available.

Buy from a reputable source, from refrigerated cabinet. Check 'use by' date. Look for Quality British Chicken mark on packs. This ensures high standards of rearing and production. Skin should be lightly coloured, slightly moist and unbroken with no dark patches. Breast meat should be plump and skin undamaged. Limbs should be fresh with undamaged skin.

Refrigerate directly after purchase at 0-5°C. Store loosely covered in greaseproof paper or foil. **Fresh uncooked** Keep in bottom of refrigerator for up to 2 days. **Cooked** Keep towards top of refrigerator for up to 2 days. **Thawed** Do not keep in refrigerator for more than 2 days. **Stuffing** Remove and refrigerate separately in a covered container. Use within 2 days. **Giblets** As stuffing. **Giblet stock or gravy** Refrigerate for up to 2 days.

Freeze directly after purchase at –18°C or below. Cooked drumsticks, thighs and wings can be frozen on the bone; cooked breast meat removed from bone should be left whole, diced or sliced. Wrap in foil or freezer bags, seal and freeze for up to 2 months. Freeze ready-frozen whole chicken in original packaging for up to 3 months (2 months if cooked). Thaw in bottom of refrigerator for several hours or, if large, in a cool room. Thaw in wrapping, but puncture seal and stand in a container. Throw away the liquid. Make sure bird is fully defrosted (no ice crystals left in cavity, legs and thighs move easily). Thawing time depends on size. The bigger the bird, the longer the thawing time. Freeze stuffing and giblets separately. After thawing use within 2 days. Do not refreeze after thawing.

## Preparing and cooking

Chicken must be cooked thoroughly. Do not cook from frozen. Chicken should not be rinsed before cooking, just wiped with damp kitchen paper. Remove giblets. When stuffing chicken, either stuff neck end only or fill cavity no more than $2/3$ full. Do not stuff a raw chicken with a warm stuffing mixture. Always weigh after stuffing to calculate cooking time. Ready-stuffed fresh or frozen birds and chilled or frozen ready-meals: see label. To see if chicken is cooked, use a skewer to pierce thickest part of thigh deeply: if juices run clear, it is cooked. If juices are at all pink, return the chicken to the oven and cook further. Cool cooked chicken quickly, cover and store in refrigerator. Reheat cooked chicken thoroughly until piping hot. Do not reheat more than once. Cooking methods include baking, barbecuing, boiling, braising, deep-frying, en papillote, frying, grilling, marinating, microwaving, poaching, pot roasting, roasting, sautéing, spit-roasting, stewing, stir-frying.

## Typical uses

Casseroles, chicken cacciatora/chasseur, cock-a-leekie, coq au vin, cordon bleu, fricassee, chicken Kiev, mousselines, pies, pot roast, quenelles, roast chicken, salads, sandwiches, soups, stews, stir-fries.

## Alternative forms

Breaded or battered portions, burgers, canned, chicken feet, chilled, frozen and dehydrated ready-meals, livers for pâté, meat paste, nuggets, organic, sausages, smoked, vacuum-packed.

| Food | How to choose | Storage | Freezing |
|------|---------------|---------|----------|
| **Duck/duckling** Breeds include Aylesbury, Barbary, Gressingham, Nantais. Available fresh and frozen, oven-ready, as whole birds and in portions such as breasts and legs. Meat varies a great deal. High in fat compared with other poultry. | Buy from a reputable source, from refrigerated cabinet. Check 'use by' date. Choose plump, not fat, birds with light-coloured, slightly moist skin. Ensure packaging and bird are undamaged. | As chicken. | Freeze directly after purchase at −18°C or below. Cooked drumsticks, thighs and wings can be frozen on the bone; cooked breast meat removed from bone should be left whole, diced or sliced. Wrap in foil or freezer bags, seal and freeze for up to 2 months. Freeze ready-frozen whole duck in original packaging for up to 3 months (2 months if cooked). Thaw in bottom of refrigerator for several hours or, if large, in a cool room. Thaw in wrapping, but puncture seal and stand in a container. Throw away the liquid. Make sure bird is fully defrosted (no ice crystals left in cavity, legs and thighs move easily). Thawing time depends on size. The bigger the bird, the longer the thawing time. Freeze stuffing and giblets separately. After thawing, use within 2 days. Do not refreeze after thawing. |

| Preparing and cooking | Typical uses | Alternative forms |
|---|---|---|
| Do not cook from frozen. Remove giblets. Duck should not be rinsed before cooking, just wiped with kitchen paper. When stuffing, either stuff neck end only or fill cavity no more than $^2/_3$ full. Do not stuff a raw duck with a warm stuffing mixture. Always weigh after stuffing to calculate cooking time. Ready-stuffed fresh or frozen birds and chilled or frozen ready-meals: see label. Prick skin of bird all over with fork before cooking and stand on a rack or trivet whilst roasting to ensure fat drains away. Eat rare or well-cooked. To see if bird is cooked, use a fork to pierce meat. Juices of rare-cooked bird run pink; well-done meat is tender and falls easily from the fork. Cool cooked duck quickly, cover and store in refrigerator. Reheat cooked duck thoroughly until piping hot. Do not reheat more than once. Cool drained fat and store covered in the refrigerator for several weeks, for use in baking or to cook potatoes. Fat can also be frozen. Cooking methods include braising, casseroling, frying, grilling, microwaving, pan-frying, pot roasting, roasting, sautéing, stewing, stir-frying. | Duck à l'orange, casseroles, cassoulet, Chinese (e.g. Peking duck), crispy, roast, salads, soups, sweet and sour. | Canned or vacuum-packed foie gras, confit, duck eggs, duck webs, sausages, smoked. |

| Food | How to choose | Storage | Freezing |
|------|---------------|---------|----------|
| **Goose**<br>Rich, fatty, gamey bird. Available fresh and frozen, oven-ready. Many geese are bred as free-range, even though they are raised commercially. | Buy from a reputable source, from refrigerated cabinet. Check 'use by' date. Available all year round. Look for a light-coloured, fatty, slightly moist skin and plump breast. Check packaging and bird are undamaged. | As chicken. | Freeze directly after purchase at −18°C or below. Cooked drumsticks, thighs and wings can be frozen on the bone; cooked breast meat removed from bone should be left whole, diced or sliced. Wrap in foil or freezer bags, seal and freeze for up to 2 months. Freeze ready-frozen whole goose in original packaging for up to 4 months. Young birds may keep longer as they contain less fat. Thaw in bottom of refrigerator for several hours or, if large, in a cool room. Thaw in wrapping, but puncture seal and stand in a container. Throw away the liquid. Make sure bird is fully defrosted (no ice crystals left in cavity, legs and thighs move easily). Thawing time depends on size. The bigger the bird, the longer the thawing time. Freeze stuffing and giblets separately. After thawing use within 2 days. Do not refreeze after thawing. |

| Preparing and cooking | Typical uses | Alternative forms |
|---|---|---|
| Goose must be cooked thoroughly. Do not cook from frozen. Goose should not be rinsed before cooking, just wiped with damp kitchen paper. Remove giblets. When stuffing goose, either stuff neck end only or fill cavity to no more than $2/_3$ full. Do not stuff a raw goose with a warm stuffing mixture. Always weigh after stuffing to calculate cooking time. Ready-stuffed fresh or frozen birds: see label. To see if goose is cooked, use a skewer to pierce thickest part of thigh deeply: if juices run clear, it is cooked. If juices are at all pink, return the goose to the oven and cook further. Cool cooked goose quickly, cover and store in refrigerator. Reheat cooked goose thoroughly until piping hot. Do not reheat more than once. Cooking methods include braising, pot roasting, roasting, stewing. | Cassoulet, roasts, salads. | Confit, eggs, fresh, canned or vacuum-packed foie gras, livers for pâté, sausages, smoked. |

| Food | How to choose | Storage | Freezing |
| --- | --- | --- | --- |
| **Poussin** Small, succulent chicken (also known as spring chicken) weighing about 450g (1lb). One cooked poussin is sufficient to serve one person. Available fresh or frozen (air-frozen), oven-ready. | Buy from a reputable source, from refrigerated cabinet. Check 'use by' date. Look for Quality British Chicken mark on packs. This ensures high standards of rearing and production. Buy whole or spatchcocked (breastbone flattened, bird spread out) or boned out and stuffed. Half-poussins are also sometimes available. | Refrigerate directly after purchase at 0-5°C. Store loosely covered in greaseproof paper or foil. **Fresh uncooked** Keep in bottom of refrigerator for up to 2 days. **Cooked** Keep near top of refrigerator for up to 2 days. **Thawed** Refrigerate for up to 24 hrs. **Stuffing** Remove and refrigerate separately in a covered container. Use within 2 days. **Giblets** As stuffing. **Giblet stock or gravy** Refrigerate for up to 2 days. | Freeze directly after purchase at −18°C or below. Cooked drumsticks, thighs and wings can be frozen on the bone; cooked breast meat removed from bone should either be left whole, diced or sliced. Wrap in foil or freezer bags, seal and freeze for up to 2 months. Freeze ready-frozen whole poussin in original packaging for up to 3 months (2 months if cooked). Thaw in bottom of refrigerator for several hours before cooking. Thaw in wrapping, but puncture seal and stand in a container. Throw away the liquid. Make sure bird is fully defrosted (no ice crystals left in cavity, legs and thighs move easily). After thawing, use within 24 hrs. Do not refreeze after thawing. |

| Preparing and cooking | Typical uses | Alternative forms |
|---|---|---|
| Do not cook from frozen. Quick and easy to prepare and cook. First, rinse in cold water and pat dry with kitchen paper. One poussin takes approximately 20-30 mins to cook, if roasted in hot oven or grilled under medium heat. Poussin may also be barbecued, fried or stuffed before cooking. Do not stuff raw bird with warm stuffing mixture. Weigh after stuffing to calculate cooking time. Cool cooked bird quickly, cover and store in refrigerator. Reheat cooked poussin thoroughly until piping hot. Do not reheat more than once. | Grilled, marinated, roasted, served with a sauce, stuffed. | |

| Food | How to choose | Storage | Freezing |
|---|---|---|---|
| **Turkey**<br>Available all year round: whole birds include bronze/black, free-range, ready-basted, standard, fresh and frozen. Turkey portions such as breast steaks, fillets, drumsticks, thighs, breast and leg joints, wings, boned and rolled joints and escalopes are available fresh and frozen, as are diced and minced meat, pre-stuffed roasts, stir-fry strips, breaded products and chilled or frozen prepared dishes. | Buy from reputable source, from refrigerated cabinet. Check 'use by' date. Look for firm, broad, round-breasted birds with unblemished white skin and a faint bluish tinge to flesh underneath. Thighs and drumsticks should be meaty and look moist, not dry and wizened. | As chicken. | Freeze directly after purchase at −18°C or below. Cooked drumsticks, thighs and wings can be frozen on the bone; cooked breast meat removed from bone should be left whole, diced or sliced. Wrap in foil or freezer bags, seal and freeze for up to 2 months. Freeze ready-frozen whole turkey in original packaging for up to 3 months (2 months if cooked). Thaw in bottom of refrigerator for several hours or, if large, in a cool room. Thaw in wrapping, but puncture seal and stand in container. Throw away the liquid. Make sure bird is fully defrosted (no ice crystals left in cavity, legs and thighs move easily). Thawing time depends on size and can take 2-48 hrs. The bigger the bird, the longer the thawing time. Remove giblets and neck as soon as possible during thawing and use to make stock or gravy. Freeze stuffing and giblets separately. After thawing use within 2 days. Do not refreeze after thawing. |

| Preparing and cooking | Typical uses | Alternative forms |
|---|---|---|
| Turkey must be cooked thoroughly. Do not cook from frozen. Remove giblets. Wipe with damp kitchen paper. Wash inside bird thoroughly, dry with kitchen paper. When stuffing turkey, stuff just before cooking. Stuff neck end only and not body cavity. Do not stuff raw bird with warm stuffing mixture. Always weigh after stuffing to calculate cooking time. Ideally, cook the stuffing separately. After cooking, allow whole roast turkey 15 mins' standing time before carving. Do not cook whole turkey in a microwave oven – bird is too big to cook evenly and thoroughly. To see if turkey is cooked, use a skewer to pierce thickest part of thigh deeply: if juices run clear, it is cooked. If juices are at all pink, return turkey to the oven and cook further. Cool cooked turkey quickly, cover and store in refrigerator. Reheat cooked turkey thoroughly until piping hot. Do not reheat more than once. Cooking methods include baking, barbecuing, boiling, braising, casseroling, frying, grilling, marinating, poaching, roasting, sautéing, stir-frying. | Burgers, casseroles, chilli, curries, in place of red meat in dishes such as spaghetti bolognese, kebabs, marinated, moussaka, mousse, pasties, pâté, rissoles, roast joint, salads, soups, stews, stir-fries. | Burgers, canned, chilled, frozen and dehydrated ready-meals, livers for pâté, organic, sausages, smoked, turkey rashers. |

| Food | How to choose | Storage | Freezing |
|------|---------------|---------|----------|

**MEAT AND POULTRY: GAME**

**Feathered game**
Examples: grouse, guinea-fowl, partridge, pheasant, pigeon, quail, snipe, squab, wild duck, woodcock. Characteristic texture and flavour depending on type: generally darker, stronger in flavour and sometimes tougher than poultry. Game birds can be bought ready-hung, fresh in the feather or ready-plucked and drawn. Some such as pheasant and quail are available oven-ready, frozen or fresh. Stuffed oven-ready quails and packets of fresh mixed diced game are also available. Hunting seasons may affect supplies.

Check 'use by' date. Important to know age of bird, which will determine how to cook it. Look for soft, plump breast meat, clean soft-textured feet (not scarred or calloused), flexible breastbones and short rounded spurs, all of which indicate youth. Avoid badly damaged birds.

Hang in cool, dry, airy place before plucking and cleaning, for anything from several days to a few weeks depending on type. This tenderises the meat and develops the flavour. Bought game should be correctly prepared, hung and drawn; ask the supplier. Store prepared oven-ready game in bottom of refrigerator at 0-5°C for up to 2 days. Use by recommended date.

Freeze directly after purchase at −18°C or below. The fattier the bird, the shorter its freezer life. Freeze with or without feathers, but only after drawing and cleaning. Cooked drumsticks, thighs and wings can be frozen on the bone; cooked breast meat removed from bone should be left whole, diced or sliced. Wrap in foil or freezer bags, seal and freeze for up to 2 months. Freeze ready-frozen game in original packaging for up to 3 months (2 months if cooked). Thaw in bottom of refrigerator for several hours or, if large, in a cool room. Thaw in wrapping, but puncture seal and stand in a container. Throw away the liquid. Make sure bird is fully defrosted (no ice crystals left in cavity, legs and thighs move easily). Thawing time depends on size. The bigger the bird, the longer the thawing time. Freeze stuffing and giblets separately. Do not refreeze after thawing.

| Preparing and cooking | Typical uses | Alternative forms |
| --- | --- | --- |
| Do not cook from frozen. Game should not be rinsed before cooking, just wiped with damp kitchen paper. Remove giblets. When stuffing bird, either stuff neck end only or fill cavity to no more than $2/_3$ full. Do not stuff a raw bird with a warm stuffing mixture. Always weigh after stuffing to calculate cooking time. Ready-stuffed fresh or frozen birds: see label. Cooking time depends on bird. Serve rare or well done, as preferred. For roasting, bard with fat. Game cooked without hanging will have an uncharacteristic mild flavour. To see if bird is cooked, use a fork to pierce meat. Juices of rare-cooked bird run pink; well-done meat is tender and falls easily from the fork. Cool cooked bird quickly, cover and store in refrigerator. Reheat cooked meat thoroughly until piping hot. Do not reheat more than once. Cooking methods include barbecuing, braising, casseroling, grilling, pot-roasting, roasting, sautéing, stewing. | Casseroles, en papillote, game pie, marinated, pâtés, pies, potted pheasant, roasts, stews, stuffed and roasted, terrines. | Canned, guinea-fowl eggs, partridge eggs, pheasant eggs, pigeon eggs, quails' eggs, smoked. |

| Food | How to choose | Storage | Freezing |
|------|---------------|---------|----------|
| **Furred game** Examples: bear, buffalo, deer (venison), elk, hare, rabbit, squirrel. Characteristic texture and flavour; generally darker, stronger in flavour and sometimes tougher than that of beef, lamb, pork. Some game, such as rabbit and venison, is available ready-frozen. | Availability is sometimes restricted by hunting seasons. Check 'use by' date. Look for plump game with perfect fur coats (if sold whole, as rabbits and hares). Avoid game with fur that is matted, dull or patchy. Young hares and rabbits should have soft, thin ears which tear easily and small white teeth. Should smell fresh. Sold whole with fur (rabbits and hares), jointed (rabbits and hares), or in cuts (as with venison); cuts are similar to cuts of lamb. The darker the flesh of the animal, the more mature it is. | Hang in cool, dry, airy place before skinning and cleaning, for anything from several days to a few weeks depending on type. This tenderises the meat and develops the flavour. Rabbits and hares are skinned and paunched after hanging. Bought game should be correctly prepared, hung and drawn; ask the supplier. Store prepared oven-ready game in bottom of refrigerator at 0-5°C for up to 2 days. Use by recommended date. | Freeze directly after purchase at −18°C or below. Defrost frozen meat thoroughly. Thaw in bottom of refrigerator for several hours or, if large, in a cool room. Thaw in wrapping, but puncture seal and stand in a container. Throw away the liquid. Make sure meat is fully defrosted. Thawing time depends on the size of the animal. The bigger the animal, the longer the thawing time. Freeze stuffing and offal separately. Do not refreeze after thawing. **Cooked meat dishes** Freeze in airtight container for up to 3 months. **Fresh game** Wipe with absorbent kitchen paper to remove excess moisture, pack and freeze. Cover with foil or clingwrap, or seal in a freezer bag or airtight container before freezing. Wrap individual portions separately if desired. **Mince** Freeze for up to 3 months. **Rabbit and hare** Treat as poultry. **Ready-frozen** Freeze in original packaging for up to 3 months (2 months if cooked). **Venison** Freeze directly after purchase. Treat as beef. **Whole joints or pieces** Wrap as fresh game, freeze for 9-12 months. |

| Preparing and cooking | Typical uses | Alternative forms |
|---|---|---|
| Do not cook from frozen. Eat rare or well done, as preferred. To see if meat is cooked, pierce with fork. Juices of rare-cooked meat run pink; well-done meat is tender and no longer pink. Cool cooked meat quickly, cover and store in refrigerator. Reheat cooked meat thoroughly until piping hot. Do not reheat more than once. Venison can substitute for beef in recipes. Cooking methods include barbecuing, braising, casseroling, grilling, marinating, pan-frying, roasting, sautéing, stewing. | Casseroles, game pie, jugged hare, marinated, pasta sauces, pâtés, roasts, soups, stews, terrines. | Canned, dried venison, minced venison hamburgers, potted venison, smoked venison, game sausages. Hares' blood is sometimes sold in cartons. |

| Food | How to choose | Storage | Freezing |
|------|---------------|---------|----------|

**MEAT AND POULTRY: SPECIALITY MEATS**

| Food | How to choose | Storage | Freezing |
|------|---------------|---------|----------|
| Unusual and exotic meats include alligator, bison, crocodile, emu, kangaroo, kid, ostrich, peacock, wild boar. Meat is usually from extensively farmed wild animals. They mature slowly, producing exceptionally lean, flavoursome meat. Crickets and locusts have also joined the list of exotica. | Buy from a reputable source, from refrigerated cabinet. Check 'use by' date. Meat should be well hung: length of time for hanging depends on type. **Boar and bison** 2 or more weeks. **Peacock and kid** 1 or more weeks. **Kangaroo, ostrich and emu** Under 1 week. Most meat hung correctly is deep ruby-red in colour with pleasant aroma. Should not look dry. Pale-coloured meats include alligator and crocodile. Check packaging is undamaged, particularly if meat was bought via mail order. Vacuum packaging is recommended for frozen meat (avoids freezer burn). If possible, verify origin of meat. Purchase wild boar from a BWBA (British Wild Boar Association) registered herd. | Refrigerate directly after purchase at 0-5°C. Use by recommended date. **Fresh uncooked meat** Store towards bottom of refrigerator for up to 7 days. Place unpackaged meat (i.e. not vacuum-packed) on tray and cover with greaseproof paper, allowing air to circulate. **Uncooked sausages** Refrigerate for up to 5 days. **Cooked joint** Refrigerate for up to 3 days. **Cooked sliced meat** Refrigerate for up to 2 days. | Freeze directly after purchase at −18°C or below. Cover with foil or clingwrap, seal in freezer bag or airtight container and freeze for up to 1 year. Taste deteriorates over time. Defrost thoroughly. Do not refreeze after thawing. |

**NUTS**

| Food | How to choose | Storage | Freezing |
|------|---------------|---------|----------|
| Seed or fruit comprising an edible kernel surrounded by a hard outer shell. Types include almonds, brazil nuts, cashew nuts, chestnuts, coconuts, hazelnuts, macadamia nuts, peanuts, pecan nuts, pine nuts or kernels, pistachios, walnuts. Flavours vary. | Check 'best before' date. Buy from a shop that has a fast turnover. Available whole in shell, or shelled or processed into flakes, halves, split or slivered, chopped pieces, ground, nibbed, blanched. Whole nuts should feel heavy for their size, be plump, have uniform colour and smell fresh. Discard damaged nuts with any signs of mould or cracked shells. | Nuts keep well and may be stored in the refrigerator at 0-5°C. Nuts in their shells keep longer than shelled nuts; shelled whole nuts keep longer than processed nuts. Nuts, particularly shelled/processed nuts, easily lose their flavour and turn rancid due to their high oil content. Exposure to heat, moisture or light will reduce their shelf-life. Store whole nuts in a cool, dry place for up to 3 months. After opening, store in airtight container in a cool, dark, dry place for a few weeks only. Use by recommended date. | Freeze whole nuts in their shells at −18°C or below for up to 6 months. Freeze shelled nuts (whole or processed) for up to 6 months. |

| Preparing and cooking | Typical uses | Alternative forms |
|---|---|---|
| Cooking time depends on cut and type of meat. Use sharp knife when carving. Cooking methods include baking, braising, casseroling, frying, grilling, roasting, sautéing, stewing, stir-frying. | Good for special occasions, dinner parties, barbecues etc. Casseroles, goulash, jambalaya, kebabs, meatballs, pasta dishes, rack of kid, salads, stir-fries, stuffed. | Boar sausages, bison sausages, kid sausages, boar patties, bison patties. |
| Widely used in sweet and savoury dishes. Nuts can be lightly toasted to enhance flavour. Use good, strong nutcrackers to crack open whole nuts. Do not apply too much pressure otherwise both shell and nut will be crushed. Nuts that have become stale or soft may be crisped up again by being roasted in the oven for a short period of time. | In baking (cakes, biscuits, puddings), Bakewell tart, baklava, burgers, Chinese, Indian and Mediterranean dishes, confectionery, decoration, desserts, ice-cream, macaroons, marzipan, nut roasts, pasties, pecan pie, pesto, salads, satay sauce, stir-fries, Waldorf salad. | Candied, canned chestnuts (whole or puréed), coconut milk, creamed coconut, desiccated or toasted coconut, dried coconut, essences such as almond, jam, ketchup, nougat, nut butter or paste, nut oils (e.g. walnut and hazelnut), peanut butter, pesto sauce, pickled walnuts, praline, pre-packed ready-made golden and white marzipan, snacks (plain, salted, toasted, natural, coated with honey, sugar, chocolate or yoghurt), spiced. |

| Food | How to choose | Storage | Freezing |
|------|---------------|---------|----------|

## PASTA AND NOODLES

Available fresh and dried, made with or without egg, in many shapes, sizes, colours and flavours; examples include cannelloni, capellini, conchiglie, farfalle, fettucine, fusilli, gnocchi, lasagne, linguine, macaroni, orecchiette, penne, ravioli, spaghetti, spätzle, tagliatelle, tortellini, vermicelli. Flavoured pasta includes asparagus, basil, black olive, black squid, chilli, corn and parsley, garlic and tomato, porcini, smoked salmon, spinach, tomato. Fresh and dried filled pasta is also available.

**Fresh** Buy from a refrigerated cabinet. Refrigerate as soon as possible after purchase. Check 'use by' date. **Dried** Avoid damaged, split or open packets. Check 'best before' date

**Fresh** Keep chilled in the refrigerator at 0-5°C. **Dried** Store unopened in packet in cool, dry cupboard. Opened: keep in airtight container as above. Keeps well for many months. **Cooked pasta** Cool quickly, cover tightly and keep in the refrigerator at 0-5°C. Plain cooked pasta reheats well and quickly in the microwave. Reheat cooked pasta/pasta dishes thoroughly until piping hot. Never reheat more than once. Use by recommended date.

Store in freezer at −18°C or below according to suitability. **Fresh** Freeze uncooked fresh pasta for up to 1 month. Can be cooked from frozen: add a couple of mins' extra cooking time. **Dried** Not applicable. **Cooked pasta** Dishes freeze reasonably well, although on thawing pasta is soft and breaks up more easily. Some cooked pasta dishes such as lasagne and cannelloni freeze very well for up to 3 months.

| Preparing and cooking | Typical uses | Alternative forms |
| --- | --- | --- |

Allow 115-225g (4-8oz) pasta per person for main course, slightly less, 55-115g (2-4oz), per person for a starter. Cook pasta in a large pan of boiling water, add salt and a little oil to the cooking water to prevent the water from frothing up and boiling over. Fresh pasta takes about 3 mins to cook. Dried pasta takes slightly longer to cook (about 10-12 mins depending on type). Some take less time than others to cook: see label. Filled pasta, both fresh and dried, generally takes slightly longer to cook: see label. When cooked, pasta should be *al dente* – tender but firm, not soft. Drain pasta as soon as it is cooked and serve immediately. Cook dried noodles in boiling water for about 3 mins or less: see label. Some noodles are just soaked in boiling water before serving.

Many dishes, usually savoury, including cannelloni, chow-mein, lasagne, macaroni cheese, pasta and sauce, pasta salad, ravioli, spaghetti bolognese. Can also be boiled and then baked into e.g. a ring, mould or loaf. In Italy pasta is sometimes served filled with fruit or as a dessert.

Barley pasta, boil-in-the-bag, brown rice noodles, buckwheat noodles, buckwheat pasta, canned pasta in sauce, packaged ready-meals, corn pasta, dehydrated pasta meals, flavoured noodles, frozen cooked pasta, instant noodles, low-protein pasta, organic, quick-cook, rice pasta, spelt pasta, wholewheat pasta.

| Food | How to choose | Storage | Freezing |
|------|---------------|---------|----------|
| **PASTRY** | | | |
| Pastry is dough made from flour, water and usually fat, sometimes incorporating other ingredients such as eggs or sugar. Available chilled, ready-made or frozen e.g. filo, flaky, puff, shortcrust, sweet dessert pastry. Fresh: filo (in sheets), puff, shortcrust. Frozen: filo, puff, shortcrust (in block or sheets), vol-au-vent cases, sweet dessert pastry. Home-made pastry includes choux, filo, flaky, flavoured shortcrust, hot-water crust, pâte sucrée, puff, rough puff, shortcrust, sweet, wholemeal. | Buy from refrigerated or freezer cabinets. Check 'best before' date. Avoid damaged or split packets. | Refrigerate directly after purchase at 0-5°C. Chill in the refrigerator for several days (see label). **Home-made uncooked pastry** Wrap in greaseproof paper and foil and refrigerate for up to 3 days. **Cooked or filled pastry** Depends on filling: see label. **Raw choux pastry dough** Refrigerate for up to 12 hours. Use bought pastry by recommended date. | Freeze in original packaging at –18°C or below for several months (see label). Pack individual pastries, cases or shapes in rigid containers to avoid damage. **Ready-frozen** See label. **Home-made uncooked pastry** Wrap well and freeze for up to 3 months. Thaw in refrigerator for several hours before using. **Cooked or filled pastry** Depends on filling: see label. **Raw choux pastry** Pipe dough into shapes, open-freeze, then pack in freezer bags and seal once frozen. Freeze for up to 3 months and cook from frozen. |
| **SAUCES AND PICKLES** | | | |
| Primarily used to accompany sweet and savoury dishes. Many different types available including chutneys, jellies, pickles, relishes, sauces, spreads. Chutneys, pickles and sauces are often home-made. | Check 'best before' date. Avoid opened or damaged jars or containers. Choose clean, tightly sealed jars, bottles or containers. | **Unopened** Store in cool, dark, dry cupboard. **Opened** Many types need to be stored in the refrigerator at 0-5°C after opening, others should be stored as above: see label. Most will keep indefinitely if stored correctly, but may darken and lose flavour with age. Reseal jars and bottles tightly to prevent product drying out. Use by recommended date. | Not applicable. |

## Preparing and cooking | Typical uses | Alternative forms

Always cooked. Most commonly baked and served hot, warm or cold. Some pastry doughs are deep-fried e.g. for spring rolls.

For sweet and savoury dishes including apple strudel, choux buns, eclairs, en croûte, gougère, mille-feuilles, palmiers, pastries, pork or game pie, profiteroles, pies, tartlets, tarts and puddings (e.g. apple, treacle, steak and kidney), turnovers, vol-au-vents.

Shortcrust pastry packet mixes. Ready-made chilled or frozen pastry items and dishes, both sweet and savoury.

Some sauces such as tomato ketchup, soy sauce, oyster sauce etc. are added to sweet and savoury dishes to add flavour, colour and texture. Some are used for basting meats, poultry and fish, and for marinades, sauces, gravies. Some, such as Worcestershire sauce and chilli sauce, have strong characteristic flavours so should be used sparingly.

Often served as a condiment to accompany meals such as cold meat, cheese, bread, main courses and starters, sandwiches. Add sauce to dishes such as casseroles, marinades, stews and stir-fries.

Organic sauces and pickles.

111

| Food | How to choose | Storage | Freezing |
|------|---------------|---------|----------|
| **SUGAR, PRESERVES AND CHOCOLATE** | | | |
| **Sugar**<br>Produced from sugar beet and sugar cane. Available in various forms including raw, refined, brown, cubes, crystals and lumps, flavoured or coloured. Many types including caster, demerara, granulated, icing, molasses, muscovado, preserving and soft brown. | Avoid split, damaged or open packs. | Keeps indefinitely in airtight container in cool, dry place. Dries quickly and forms lumps or a block when exposed to air. To soften hard sugar, warm in a low oven. Sugar syrup, unrefrigerated, will keep for several days; refrigerated, for several weeks. | Freeze in prepared dishes such as cakes, puddings, biscuits. |
| **Syrups**<br>Examples: corn, glucose, golden, malt extract, maple, molasses, treacle. All are liquid forms of sugar. Colours and flavours vary: darker syrups have more distinct, lighter syrups more delicate flavour. | Check 'best before' date. Avoid damaged, split or open packaging. | Keep indefinitely if stored in dry place at room temperature. Maple syrup keeps indefinitely, ideally in refrigerator. If syrup crystallises, place container in a saucepan of warm water until it liquefies. Use by recommended date. | Freeze in dishes. |
| **Sugar substitutes (sweeteners)**<br>Alternatives to sugar, for those on low-calorie diets. Types include saccharine and aspartame (otherwise known as Nutrasweet). They are much sweeter than sugar and should be used in small amounts. Saccharine is about 300 times sweeter than sugar. | Available as a liquid, granulated, powder or capsules/tablets. Check 'best before' date. | Store at room temperature in cool, dry place in sealed package or airtight container. Use by recommended date. | Can be frozen in dishes. |

| Preparing and cooking | Typical uses | Alternative forms |
|---|---|---|
| Sugar syrup is made by boiling mixture of sugar and water to varying temperatures, depending on the strength and type of syrup required. Use for fruit salads, to moisten cakes, glaze pastries, poach and glaze fruit, as basis of sorbet, to make sweets and confectionery. Syrup changes composition as it boils and goes through various stages from thread to hard-crack. Always sift icing sugar before use. In many dishes brown sugar can be substituted for white. | Use as a sweetener, also as preservative, flavouring or tenderiser; adds texture. In cakes, puddings, desserts, confectionery such as toffee, gingerbread, chutney, preserves, biscuits, with fruit, glazes, sauces, hot beverages, pastries, meringues, decorations, icings. | Fructose, glucose, glycerine, jaggery, date sugar, brown palm sugar, vanilla, ready-to-roll plain and coloured icings, icing for cake decoration sold in tubes, icing kits, chocolate-flavoured icing sugar. |
| Heat syrups gently to avoid burning. | In baking, such as cakes, biscuits, fruit cakes, sauces, sweets, desserts and puddings, toppings for desserts and fruit, gingerbread, parkin, flapjacks, steamed pudding, topping for ice-cream, pancakes, waffles, treacle tart, glazing gammon or ham. | Flavoured syrups such as chocolate or raspberry. |
| Some can be used in cooking: see label. Sprinkle, spoon over or stir into foods. | For sweetening drinks and cold desserts. Used in a wide range of low-calorie products, including yoghurts and desserts. | |

| Food | How to choose | Storage | Freezing |
|------|---------------|---------|----------|
| **Honey**<br>Oldest sweetening substance, produced by bees from nectar of flowers and trees. Many varieties. Flavour and appearance are determined by flower. Appearance ranges from light and clear to set or thick and opaque. Usually, the paler the honey, the milder the flavour. Each type has its own distinctive flavour, varying from mild to aromatic and flavoursome. Honey improves keeping quality of baked goods. | Check 'best before' date. Avoid damaged packaging. | Keep in sealed container in cool, dry place. Lasts a long time if stored properly. Do not refrigerate: many honeys crystallise and become grainy if too cold. Honey also tends to crystallise and harden with age. If it crystallises, place the jar in a saucepan of warm water until it liquefies. Use by recommended date. | Can be frozen in dishes. |
| **Preserves**<br>Examples: jams, jellies, conserves, marmalades, fruit butter, cheeses and curds. Combination of fruit and sugar: varieties include single fruit or fruit combined with other fruits or flavourings, such as herbs. Many fruits are suitable for preserving. | Check 'best before' date. Avoid damaged packaging. | Well-sealed, home-made preserves keep for well over 1 year in a cool, dark, dry place. In time flavour fades and quality deteriorates. Use bought preserves by recommended date. | Can be frozen in dishes. |
| **Carob**<br>Flavouring produced from pulp of the carob or locust bean. Often substituted for chocolate in recipes, has lower fat content and contains no caffeine. Useful for those allergic to dairy produce. Available in plain and no-added-sugar varieties as raw powder, blocks, chips, or roasted like coffee beans. | Avoid damaged, split or open packaging. Check 'best before' date. | **Unopened** Store in a cool, dry cupboard. **Opened** Store in airtight container as above. Use by recommended date. | Can be frozen in dishes. |

| **Preparing and cooking** | **Typical uses** | **Alternative forms** |
|---|---|---|
| The more liquid the honey, the more easily it will mix with other foods or ingredients. Clear honey is usually best for cooking. Honey becomes runny when heated, which makes it easier to blend with other foods and dissolve more quickly. When using honey in place of sugar, treat as liquid and reduce other liquid ingredients accordingly. Honey is sweeter than sugar so less is needed. Heat gently to avoid burning. | Use on its own or as an ingredient. Sweetens, preserves and enhances flavour. Adds moistness to baked goods such as cakes, pastries (baklava), gives texture to breadcrusts. Serve with cereals, ice-cream, poached fruits, toast, scones, yoghurt; use to glaze meat such as ham; use for barbecue sauces, chicken, coating breakfast cereals, drinks, marinades, salad dressings, vinaigrette. | Honeycomb or comb honey, squeezable honey. |
| Heat jams or preserves gently to avoid scorching. | Spread on bread, crackers, toast etc. Add to yoghurt, use as a cake filling, as a glaze for pies, tarts and meat, as a sweetener, piped for decoration. | Pure fruit spreads, organic jams and marmalades, diabetic, reduced-sugar, no-added sugar, extra jam. |
| Carob powder may be used as a substitute for cocoa powder. It has a different flavour and is much sweeter. Sieve before use. | Soft drinks, baking, confectionery, desserts, puddings, ice-cream, snack bars, on its own as a 'treat'. Can be melted for coating, although carob does not have glossy, shiny appearance of chocolate. | Carob-coated nuts, snack bars, etc. Ready-baked carob goods, e.g. biscuits, flapjacks, cakes. |

| Food | How to choose | Storage | Freezing |
|------|---------------|---------|----------|
| **Chocolate, cocoa powder and drinking chocolate** Includes unsweetened, bittersweet, bitter, plain, milk, white, sweet, couverture. Used as a drink, snack, sweet, flavouring, garnish, decoration. Combines well with many flavours including nuts, spices and mint. Taste depends on type and quality. Cocoa powder, made from ground roasted cocoa beans, is unsweetened. Drinking chocolate, made from ready-prepared mixture of cocoa powder and sugar, sometimes includes other flavourings, e.g. orange. | Depends on individual taste and intended use. Sweetness varies, as does cocoa butter or cocoa fat content: the higher this is, the richer and smoother the chocolate. Plain chocolate is best for eating and cooking. Check 'best before' date. | Do not refrigerate: low temperatures and moist air cause an unsightly greyish-white sugar bloom. This does not affect taste. Use by recommended date. **Unopened** Store in cool, dry, airy place at room temperature. **Opened** Wrap in foil or clingwrap and keep away from strong-smelling foods. | Can be frozen in dishes. |

| Preparing and cooking | Typical uses | Alternative forms |
|---|---|---|
| Plain chocolate is best for cooking: some milk chocolate is not suitable. White chocolate is sensitive to heat and difficult to handle. Use only grated or melted chocolate for cooking. Chocolate burns very easily: melt over *bain-marie* or double boiler/saucepan. Microwave ovens are ideal for melting chocolate. If overheated, chocolate becomes hard and granular. If water gets into chocolate, it will stiffen and solidify. Butter and oil add richness to melted chocolate. When combining melted chocolate with another liquid, both should be the same temperature. Chill before grating and keep as cool as possible. | Used primarily as flavouring for sweet (sometimes also savoury) dishes. In drinks such as hot chocolate, fondues; for dipping/coating fruit and nuts; in cakes, puddings, icing, sauces, gâteaux, tarts, batters; for decoration, garnish, piping, coating. | Chocolate chips, shavings, confectionery, cocoa butter, compound or coating chocolate, chocolate-flavoured icing, chocolate cake coverings (milk, plain), low-calorie or low-fat drinking chocolate, organic, low-fat chocolate. |

| Food | How to choose | Storage | Freezing |
|------|---------------|---------|----------|
| **VEGETABLES** <br> **Fruit vegetables/ vegetable fruits** <br> Examples: aubergine, avocado, mushroom, pepper family, squash family, tomato. All are fruits of their plants. | Check 'display until' date (if shown). **Aubergine** Should be firm with bright, shiny, unwrinkled, unblemished skins. **Avocado** Many varieties including knobbly- and smooth-skinned ones. When ripe, should feel soft when gently pressed at stalk end; too soft means over-ripe. Ripen in warm place. **Mushrooms** Two categories, cultivated (button, cup, field, flat) and wild (boletus, ceps, chanterelles, cloud ear, morels, oyster, shiitake). Cultivated should be white or light tan without discoloration, firm and moist with no damp patches or dry stalk ends. Depending on the variety, wild should smell fragrant and be moist, fleshy, smooth and soft with silky caps. Avoid limp mushrooms or those with dark patches. **Pepper family** Choose bright, shiny peppers and chillies with unwrinkled, unblemished skin without soft patches. **Squash family** Includes courgette, marrow, cucumber, squashes and pumpkins. Choose firm vegetables with unbroken skin, without soft or brown patches. As size increases, flavour decreases. **Tomatoes** Redness generally signifies ripeness, although yellow tomatoes are also sweet. Choose bright, firm, fresh tomatoes with tight, unwrinkled, unblemished skins. Avoid soft tomatoes. | Store in refrigerator, low down or in salad drawer at 0-5°C. **Aubergine** Wrap loosely and refrigerate for up to 3 days. **Avocado** Refrigerate, if ripe, for up to 3 days. **Mushrooms** Refrigerate in punnet or paper bag for up to 4 days. **Pepper family** Wrap loosely and refrigerate for up to 5 days. **Squash family** Wrap loosely and refrigerate for up to 4 days. **Tomatoes** Ripen at room temperature. Refrigerate for up to 3 days when ripe, longer when unripe. Bring to room temperature before eating. | **Aubergine** Freeze in dishes. **Avocado** Whole avocadoes are not suitable for freezing. Mash flesh with lemon juice and freeze for up to 4 weeks. **Mushrooms** Freeze in dishes. **Pepper family** Whole peppers not suitable for freezing. Blanch chopped peppers and freeze for up to 6 months. **Squash family** Freeze in dishes. **Tomatoes** Whole tomatoes collapse when frozen. Purées and sauces freeze well for up to 1 year. |

## Preparing and cooking

**Aubergine** Rinse and remove stalk. Bake, grill, sauté, stir-fry, stew or stuff. **Avocado** Peel and remove stone before eating. Flesh discolours quickly when cut. To avoid this, rub with lemon juice. **Mushrooms** Most do not need peeling; just wipe to remove dirt or dust immediately prior to use. Can be eaten raw or cooked for short period of time. **Pepper family** Always core and deseed peppers. Eat raw or cooked. Wear rubber or plastic gloves when handling chillies as they can irritate the skin. Leave tiny seeds in chillies if liked (they are the pungent part and add extra heat). Cooking methods include baking, deep-frying, frying, grilling, steaming, stewing, stir-frying. **Squash family** Generally, summer varieties do not need peeling. Winter varieties do because of their tough outer skin. Discard large seeds and stalks before use. Cooking methods include baking, deep-frying, frying, grilling, sautéing, stewing, stir-frying, stuffing. **Tomatoes** Eat baked, chopped and simmered, as chutney, grilled, as pizza, raw, as relish, as salad, sautéed, as soup, stuffed.

## Typical uses

**Aubergine** Au gratin, grilled, moussaka, ratatouille, stews. **Avocado** Eat as fruit or vegetable, in guacamole, salads, sauce, soup, stuffed. **Mushrooms** Casseroles, fritters, pies, salads, soup, stews, stir-fry, stuffed. **Pepper family** Grilled, marinated, pizzas, ratatouille, salads, stews, stuffed. **Squash family** Eat raw (courgette, cucumber), baked, pumpkin soup and pie, salads, stews, stir-fry, stuffed. **Tomatoes** In pasta dishes such as spaghetti bolognese, salads, sauces, stuffed.

## Alternative forms

Canned squash, canned tomatoes, courgette flowers, dried and canned mushrooms, dried tomatoes and peppers, frozen, passata, pickled or salted cucumber, pickled aubergine, prepared chilled vegetables, sun-dried tomatoes, sun-dried tomatoes in oil, tomato juice, tomato paste, tomato purée.

119

| Food | How to choose | Storage | Freezing |
|------|---------------|---------|----------|
| **Greens**<br>Examples: broccoli, Brussels sprouts, cabbage, cauliflower, chicory, Chinese leaves, kale, seakale beet, sorrel, spinach, Swiss chard, vine leaves. | Check 'display until' date (if shown). Generally, avoid limp or discoloured leaves and brown or damaged patches. **Broccoli and cauliflower** Look for firm, compact florets, strong, firm stalks and bright green leaves. Avoid limp, yellow or discoloured leaves and heads. **Cabbage/Brussels sprouts** Should be brightly coloured and feel solid with crisp, tight leaves. Old and stale cabbages and Brussels sprouts smell strong. **Spinach and leaves** (e.g. chicory, Chinese leaves) Choose small, young, tender leaves, fresh and bright in colour. Avoid thick, woody stalks. | **Broccoli and cauliflower** Store loosely covered in open polythene bag in refrigerator, low down or in salad drawer for up to 4 days. **Cabbage** Store in refrigerator, low down or in salad drawer for up to 10 days. **Brussels sprouts** Store in refrigerator, low down or in salad drawer for up to 4 days. **Spinach and leaves** Store loosely covered in an open polythene bag in refrigerator, low down or in salad drawer for up to 5 days. | All can be frozen in dishes. **Broccoli and cauliflower** Blanch for 3-5 mins, cool, open-freeze, then pack and freeze for up to 1 year. **Cabbage** Use crisp, young cabbage only. Blanch for 1-2 mins, cool, pack and freeze for up to 1 year. **Brussels sprouts** Blanch for 2-4 mins, cool, pack and freeze for up to 1 year. **Spinach** Cooked spinach freezes well. Blanch for 2 mins, press out moisture, cool, pack and freeze for up to 1 year **Leaves** Not suitable for freezing. |
| **Peas, beans and sweetcorn**<br>Members of legume family with double-seamed pods containing a single row of seeds. Fresh varieties include broad bean, butter bean, flageolet bean, French bean, green bean, haricot vert, mangetout, okra, peas, runner bean, sugar-snap peas, sweetcorn. Eaten young in pod, or as shelled peas or beans (matured and eaten without pod). Dried: see Beans and Pulses. | Check 'display until' date (if shown). Choose small, tender, young pods, which are crisp and bright in colour. Avoid wilted, damaged or browning vegetables and those with soft spots or wrinkled pods. **Sweetcorn** Choose moist, bright husks with fresh tassels and uniform, small, tightly packed kernels. | Store loosely wrapped in an open polythene or paper bag in refrigerator, low down or in salad drawer for up to 4 days. **Sweetcorn** Eat within 24 hrs of purchase. | All can be frozen in dishes. **Broad bean** Blanch for 3 mins, cool, pack and freeze for up to 1 year. **French bean** Blanch for 2-3 mins, cool, pack and freeze for up to 1 year. **Mangetout** Blanch for 1 minute, cool, dry, pack and freeze for up to 1 year. **Peas** Blanch for 1-2 mins, cool, dry, pack and freeze for up to 1 year. **Runner bean** Blanch for 2 mins, cool, dry, pack and freeze for up to 1 year. **Sweetcorn** Blanch cobs for 4-6 mins, cool and pack singly or in pairs. Freeze for up to 1 year. |

| Preparing and cooking | Typical uses | Alternative forms |
|---|---|---|
| Wash and drain leaves such as spinach thoroughly before using. Usually boiled or steamed before serving. Other cooking methods include baking, braising, sautéing, steaming, stewing, stir-frying. | As vegetable, gratin dishes, florentine (i.e. containing spinach) dishes, salads, sauces, sauerkraut, soups, stir-fries, stuffed leaves, with a sauce (cauliflower cheese). | Canned, frozen, packed in brine (vine leaves), pickled cabbage, pickled cauliflower, prepared chilled vegetables. |
| Usually boiled or steamed before serving. Some types can be eaten raw or incorporated into dishes before cooking. Generally, top and tail beans before cooking or eating. Cooking times depend on age and type. | As vegetable, or as bean hotpots, burgers, casseroles, corn chowder, salads, soups, stews, stir-fries. | Baby corn, canned, frozen, in a jar, pickled, prepared chilled vegetables. |

| Food | How to choose | Storage | Freezing |
|---|---|---|---|
| **Roots and tubers (bulbs)** Vegetables which grow underground. Examples: artichoke, beetroot, carrot, celeriac, dasheen, eddo, mooli, onion family, parsnip, potato, radish, rutabago, salsify, swede, sweet potato, taro, turnip, yam. | Check 'display until' date (if shown). Choose firm, heavy, even-sized vegetables. Avoid mouldy, damaged, bruised or wilted vegetables with dry or broken skins. **Potatoes** Avoid green or sprouted potatoes. New potato skins should rub away easily. **Onion family** Should be firm with no sign of sprouting, have dry skin and no discoloration. Leeks and spring onions should be crisp and bright in colour. Avoid dry or slimy leaves. | Generally, unrefrigerated vegetables keep well. Cut off leafy tops before storing. Store in a cool, dry, dark, airy place at room temperature or in refrigerator, low down or in salad drawer at 0-5°C for up to 2 weeks. **Potatoes** Store in cool, dry, dark airy place for up to 2 weeks, away from onions and exposure to light. **Onion family** Store separately as strong flavour may taint other foods. Store whole leeks and onions in a cool, dry, dark, airy place for up to 2 weeks. Cover chopped leeks and onions and store in refrigerator, low down or in salad drawer for up to 24 hrs. Cover spring onions loosely and store in refrigerator, low down or in salad drawer for up to 3 days. | Can be frozen in dishes. **Beetroot** Blanch for 5-10 mins, cool, pack in rigid container and freeze for up to 1 year. **Carrots** Blanch for 3-5 mins, cool, pack and freeze for up to 1 year. **Celeriac** Cook until almost tender, cool, pack and freeze for up to 1 year. **Onion family** Blanch for 2 mins, cool, pack and freeze for up to 1 year. **Parsnip** Blanch for 2 mins, cool, pack and freeze for up to 1 year. **Potatoes** Uncooked potatoes do not freeze well. Freeze cooked potato products such as croquettes for up to 6 months. **Turnip** Blanch for 2½ minutes, cool, pack and freeze for up to 1 year. Alternatively, cook, mash or purée and freeze for up to 6 months. **Swede and celery** Cook, mash or purée and freeze for up to 6 months. |
| **Salad leaves** Examples: lettuce (cos, curly endive, frisée, iceberg, lambs' lettuce, little gem, lollo rosso, oak leaf, radicchio, rocket), cress, mustard and watercress. | Check 'display until' date (if shown). Choose really fresh, crisp leaves; avoid those which are wilted or browning. Look for well-formed heart. | Store loosely wrapped in a polythene bag or damp cloth in refrigerator, low down or in salad drawer for up to 3 days. Remove all plastic wrapping from pre-packed lettuce before storing. Storage time depends on type and freshness: most salad vegetables will keep for 2-14 days if stored correctly. | Freeze watercress in dishes. Otherwise, not suitable for freezing. |

| Preparing and cooking | Typical uses | Alternative forms |
|---|---|---|
| Generally, scrub, peel or wash thoroughly before cooking (peel as thinly as possible). Top, tail and peel carrots to remove any pesticide residue. Some root vegetables (e.g. potatoes) act as thickeners when cooked. Some root vegetables (e.g. potatoes and artichokes) discolour quickly after peeling or cutting. Place in water and cook immediately. Usually boiled or steamed before serving and eaten hot. Some may be eaten raw (e.g. onion family). Other cooking methods include baking, braising, deep-frying, frying, grilling, sautéing, roasting. | As vegetable, in casseroles, cock-a-leekie, croquettes, pasties, pies, potato dauphinoise, potato pancakes, purées, roasts, rösti, salads, soups, stews, stir-fries, stuffed (onions). | Canned, carrot juice, dehydrated potato products (e.g. mash), dried, freshly minced shallots, frozen and fresh potato products such as chips, frozen vegetables, onion paste and powder, onion salt, pickled beetroot, pickled mooli, pickled onions, potato croquettes, prepared chilled vegetables (e.g. potatoes, leeks), smoked shallots, vacuum-packed beetroot. |
| Remove and discard discoloured leaves. Wash and drain/dry well before use. To avoid bruising, tear leaves instead of cutting with a knife. Serve raw in salads. Some (e.g. watercress) may be cooked in dishes: methods include baking, braising and sautéing. | Salads, sandwiches, sauces, soufflés (watercress), soups. | Prepared chilled leaves and salad mixes. |

| Food | How to choose | Storage | Freezing |
|------|---------------|---------|----------|
| **Stalks, shoots and stems** Edible vegetable stalks, shoots and stems. Examples: asparagus, beansprouts, cardoon, celery, fennel, globe artichoke, kohlrabi, seakale. | Check 'display until' date (if shown). Choose firm, crisp shoots/stalks which are bright and clear in colour. Avoid brown, wizened, dry or wilting roots. Cut stalks should be crisp and dry, not withered. **Asparagus** Look for straight, plump, even stalks and tightly-budded, compact heads. | Refrigerate in open polythene bags in refrigerator, low down or in salad drawer at 0-5°C. Wrap cut stalks in clean, damp cloth, or immerse in water (change water daily). **Artichoke (globe)** Wrap loosely and refrigerate for up to 4 days. **Asparagus** Refrigerate for up to 2 days. **Beansprouts** Refrigerate and use on day of purchase. **Celery and fennel** Refrigerate in plastic sleeve or open bag for up to 2 weeks. **Seakale** Refrigerate for up to 3 days. | **Artichoke (globe)** Blanch for 8-10 mins, cool, pack and freeze for up to 1 year. **Asparagus** Freeze when very fresh. Blanch for 2-4 mins, cool, pack and freeze for up to 9 months. **Beansprouts** Freeze in dishes. **Celery** Freeze in dishes. **Fennel** Blanch for 2 mins, cool, pack and freeze for up to 1 year. **Kohlrabi** Blanch for 1½ mins, cool, pack and freeze for up to 1 year. |
| **Quorn** Meat alternative, made from a tiny plant fermented in same way as yoghurt or beer. Available ready-frozen. | Sold pre-packed. Varieties range from mince to pieces. Check 'use by' date. Avoid damaged or split packets. | Store in original packaging in refrigerator at 0-5°C. After opening, store in a sealed polythene bag or container and use within 24 hours. Use by recommended date. | Freeze Quorn on own and in dishes. Freeze on day of purchase, in original packaging or in a sealed freezer bag or container. Store for up to 3 months. Ready-frozen Quorn is also available. |
| **Seaweed** Varieties of edible seaweed include arame, dulse, hijiki, kombu (kelp), laver, nori, wakame. Available mainly dried, but sometimes fresh. Sold shredded or in sheets. | Check 'use by' date. Avoid damaged or split packages. | Store dried and fresh seaweed in cool, dry, dark, airy place. See label. Use by recommended date. | Not applicable. |

| Preparing and cooking | Typical uses | Alternative forms |
|---|---|---|
| Usually boiled or steamed before serving. Some may be eaten raw; other cooking methods include stir-frying, adding to casseroles or braising. | As vegetable, used in stir-fries, soups, in casseroles, flans, pies, stews, gratin dishes, stuffed. | Canned, frozen, prepared chilled vegetables. |
| May be eaten 'raw' as a chilled ingredient, but is usually best cooked before eating. | Ideal meat substitute e.g. in pasta sauces, stir-fries, casseroles, lasagne, curries. | Quorn burgers, sausages, fillets, ready-meals such as lasagne, pasta, tikka masala. |
| To reconstitute dried seaweed, soak in water then boil until soft. Depending on type and dish, most are simmered before serving but can also be eaten raw or deep-fried. | Commonly used in Japanese cooking. To accompany seafood, as edible wrapper (e.g. for sushi), garnish, in patties such as laver bread, salads, soups, stir-fries. | Other sea vegetables (e.g. samphire) can be eaten as a vegetable. Agar-agar seaweed is a good vegetarian substitute for gelatine. |

# CHAPTER 3

# FOOD LABELLING

MAKING SENSE of food labels is difficult. It is generally hard to sort out what is marketing hype and what is useful information. Some of the information on food packets has to be there by law, but most of it is there just to make the product look more attractive: so it is basically advertising. The information which is legally required often seems to have been made deliberately difficult to understand. Some of the advertising claims on food packets are covered by regulations but many are not, so you need to treat them with scepticism.

This chapter tells you what information you should expect to find on food labels and how to make sense of it. It also tells you about the sorts of claim you will find and how to check their accuracy.

## Information which has to be on food packets

Food labelling law ensures that most packaging carries:

(1) the name of the food
(2) a list of ingredients
(3) a 'use by' or 'best before' date
(4) any special storage or cooking instructions
(5) the name and address of the manufacturer
(6) information about certain processes used in manufacturing, such as irradiation.

There are exceptions to these rules. In particular, foods sold in cafés and restaurants, and foods sold loose, do not have to be labelled like this, so it is often impossible to know what is in them or how long they should keep. Small packets and glass bottles, including milk bottles, also do not have to carry all of this information.

Some of the information you will find on packets does not have to be there but when it is it must be given in a standard way. This is the situation with the nutrition information table, nowadays found on most packaged foods sold in the UK.

## Names of foods

You can learn something about a food from its name because the law says that the name of a product must not be misleading and there are some more specific rules about names too. In particular the name should be sufficiently precise for the consumer to distinguish the product from other products. So, for example, describing a food as 'pâté' isn't enough, because there are many different kinds, so the label should say what sort of pâté it is.

But the law on names is not particularly helpful and you can easily be misled by perfectly legal names. For example, if the name of a food contains the name of an ingredient followed by the word 'flavour', the food does not have to contain any of that ingredient. Smoky bacon-flavour crisps do not need to contain any bacon. And just because an ingredient features in the name of product it does not mean that it contains a lot of it. The main ingredient of chicken stock cubes, for example, does not have to be chicken.

The law also says that if the name does not give enough information, manufacturers should also give a fuller description elsewhere on the pack. This needs only to be somewhere on the packet, so you could easily miss it.

## Ingredients

By law manufacturers must list the ingredients in descending order of weight. So you can get some idea of the quality of a food from the ingredients list. For example, if a fruit yoghurt lists 'sugar' before 'fruit', the yoghurt is not likely to have much fruit in it. But the ingredients list doesn't give you much help with gauging food quality because the manufacturer does not have to say how much of each ingredient there is.

Only if a label makes a claim about what the food is made of, such as 'made with fresh egg yolk' or 'with extra chocolate', does the label have to show the minimum amount of that special ingredient. The European Union is in the process of bringing into force a directive which would, at least, require manufacturers to list the amount of the main ingredient in a food. Hence, for mushroom soup the label would tell you how much mushroom the soup contained. It would be even better if the label were to list the amount of all the main ingredients: a few responsible manufacturers in fact do this already.

Ingredients do not have to be listed if they are part of another ingredient which makes up less than 25 per cent of the final product. The label on a cake, for example, could say that it includes marzipan rather than list the almonds in the marzipan. This causes all sorts of difficulties. You might be allergic to an ingredient which does not appear on the label because it is an ingredient of an ingredient. Vegetarians cannot be absolutely certain from scrutinising the ingredients list that there is no meat in a particular product, and people seeking to avoid particular types of meat for religious reasons will have problems too.

Some ingredients lists seem to contain mostly things you have never heard of. Many of the names used are just technical terms for items the food manufacturer does not particularly want to highlight. For example, sucrose, glucose, dextrose, fructose, glucose syrup, corn syrup and invert sugar are all forms of sugar. Caseins, caseinates or whey are, in fact, the protein part of milk. Hydrolysed vegetable protein is protein extracted mainly from soya beans, and chemically denatured (its purpose is to add flavour to soups, canned stew and so on).

As mentioned above, foods need not carry a nutrition information table. If there is no such table, you can consult the ingredients list instead. But if you do, beware. There are many traps for the unwary. For instance, foods containing 'hydrogenated vegetable fat or oil' are not necessarily healthier than those containing animal fat. Hydrogenation involves treating vegetable oil with hydrogen to turn it solid and in the process creates trans fatty acids – nowadays considered just as bad for your heart as the saturated fatty acids found in animal fat.

Towards the end of the ingredients list you will find the additives. These are ingredients which can be added to foods but which have no nutritional value. Additives can be natural materials, man-made nature-identical substances or artificial chemicals. They are used to prevent deterioration or the growth of harmful bacteria or moulds, to modify consistency and texture and/or to add or restore colour or flavour. All additives, apart from flavourings, have to be listed in the ingredients list. They can be given as their technical name and/or by a number given to them by the European Union – the E number.

In the wake of all the bad publicity about E numbers since the 1980s manufacturers normally now give the technical names of the additives rather than their E numbers. Just because an ingredients list has no E numbers in it does not mean to say that the food is free from additives.

Additives include 'processing aids', which are chemicals used in the manufacture of foods for technical reasons: for example, to stop ingredients sticking to machinery, to help freeze foods faster, and so on. These processing aids have to be listed only when they remain in the food when it is sold. The chart on pages 131-3 lists the most common additives, including processing aids, where you will find them and what they are for.

## Are additives really necessary?

Some additives need to be in foods but many do not. Food preservatives such as vinegar (E260) have been used in the production of foods throughout history. Sodium bicarbonate (E500) has long been used for baking cakes. Some preservatives are definitely needed to slow down the growth of harmful micro-organisms such as salmonella and other bacteria which cause food poisoning.

Fruit-based squashes, for example, need preservatives, because people use them over a long period, opening and closing the bottles and exposing the contents to conditions which favour yeasts and moulds.

But not all additives are necessary. Some food retailers and manufacturers have made moves to reduce the use of additives in response to public pressure. For example, manufacturers have been able to manage without preservatives in many fruit yoghurts by improving food hygiene during manufacture and by introducing refrigerated distribution systems.

Food manufacturers also argue that additives are necessary because they help to improve texture, taste and colour. For example, without emulsifiers low-fat spreads would separate into oil and water. Flour improvers make the texture of bread and cakes lighter and polyphosphates incorporate water into cooked meats so that they seem more tender. Canned peas without added colour would appear grey and unappetising.

There are huge numbers of permitted additives, each of them designed for a different use. For example, some of the 44 different permitted colours are not heat-stable: they could not be used in products such as fruit purée or canned soup which are cooked. Anothcyanins, often used to colour things red, are stable in acidic conditions but turn blue in neutral conditions, so they are used mainly in drinks and yoghurt. Copper chlorophyll (E141) is used in jelly and lime squash because natural green colouring fades on exposure to light.

## Are additives safe to eat?

No one knows for sure. A few people react badly to some additives. For example, some children are allergic to azo dyes found in sweets and soft drinks. A few cases of adverse reactions to sulphites have also been reported: they can cause asthma attacks.

Most nutritionists argue that the risks to health from additives pale into insignificance when compared with other risks, such as the chances of developing heart disease from a diet which is high in fat, particularly saturated fat. And 'natural' foods like wheat, milk, nuts and shellfish cause many more adverse reactions. On the other hand you often find that it is fatty and sugary foods which contain the most additives.

The government and many experts are satisfied that the control and use of additives is stringent and safe. (The safety of additives is nowadays assessed by the European Commission's Scientific Committee for Food.) Other experts are less sure about additives. They say that using the results of experiments on animals for an assessment of the long-term health effects on humans is dubious. Moreover, most of the research into the safety of additives is funded by the very companies that wish to sell foods with the additives in

**Additives**

| Grouping | Common types | Where they occur | What they do |
|---|---|---|---|
| Acidity regulators | Orthophosphates (E339-E341) | Processed cheeses, coffee creamer, dessert mixes, packet soups | Stabilise and/or change the acidity or alkalinity of a food |
| | Sodium citrate (E331) | Jam | |
| Acids | Acetic (E260) | Pickles, mayonnaise, salad dressings, bread | Give sharp flavour to foods, have preservative effect. The salts of acids are used for raising agents, e.g. cream of tartar |
| | Citric (E330) | Jam, processed cheese and sweets | |
| | Orthophosphoric (E338) | Soft drinks | |
| Anti-caking agents | Calcium silicate (E552) | Icing sugar | Prevent food particles sticking together or going lumpy |
| | Magnesium carbonate (E504) | Icing sugar, salt | |
| | Calcium phosphate (E341) | Instant drink mix | |
| Anti-foaming agents | Dimethylpolysiloxane (E900) | Beer | Prevent liquids from boiling over by breaking down foams and reducing scum. Stop frothing during bottling |
| Anti-oxidants | Ascorbic acid (E300) | Fruit drinks | Stop fats from going rancid and flavours from deteriorating due to oxidation; prevent browning of fruit, vegetables and fruit juices exposed to air, extend shelf-life |
| | Butylated hydroxyanisole (BHA) (E320) | Soup mixes, cheese spread, instant potato mix | |
| | Butylated hydroxytoluene (BHT) (E321) | Instant potato mix | |
| | Lecithin (E322) | Margarines and low-fat spreads | |
| Bases | Sodium bicarbonate (E500) | Cakes, biscuits | Decrease the acidity of foods and help dissolve acidic substances, such as some colourings. Release carbon dioxide for making cakes and biscuits |
| | Sodium hydroxide (E524) | Sweets, cocoa, jam | |
| Bleaching and improving agents | Ascorbic acid (E300) Azodicarbonamide (E927) Chlorine (E925) | Bread, cakes, biscuits | Bleach and artificially speed up the maturing of flour to make bread dough easier to process |

| Grouping | Common types | Where they occur | What they do |
|---|---|---|---|
| Bulking agents | Alpha-cellulose (E460) | Slimming bread | Add bulk without calories so you feel you are eating more |
| | Polydextrose (E1200) | Reduced-calorie and slimming foods | |
| Colours | Amaranth (E123) | Fruit pie fillings, yoghurt | Make food more colourful, restore colour lost during processing or storage |
| | Caramel (E150) | Beer, dessert mixes, pickled onions, cola, gravy browning, soups, sauces | |
| | Carmoisine (E122) | Jam | |
| | Red 2G (E128) | Sausages, cooked meats | |
| | Sunset yellow (E110) | Biscuits, orange squash | |
| | Tartrazine (E102) | Soft drinks, canned peas | |
| Emulsifiers | Lecithin (E322) | Chocolate | Help oils and fats to mix with water, add smoothness and creaminess of texture, extend shelf-life of baked goods |
| | Mono- and di-glycerides of fatty acids (E471) | Margarines and low-fat spreads, bread, frozen desserts, chocolate products, ice-cream | |
| | Polyoxethylene sorbitan monostearate (E435) | Ice-cream, desserts, cake mixes | |
| Enzymes (processing aids) | Proteolytic enzymes | Meat, soft drinks | Make technical changes possible in food, prevent cloudiness in drinks, turn milk into cheese |
| | Rennet | Cheese | |
| Excipients (processing aids) | Sorbitan monostearate (E491) | Cream fillings | Act as 'carriers' for other additives to make mixture more manageable or to help mix in flavours |
| Firming and agents | Calcium chloride (E509) Calcium hydroxide (E526) Calcium malate (E352) | Canned fruit and vegetables, sweets | Make processed food firm or crisp, e.g. canned carrots or pickled cabbage |
| Flavour enhancers | Inosine 5'-disodium phosphate (E631), sodium 5'-ribonucleotide (E635), monosodium glutamate (E621) | Savoury foods, soups, sauces and meat products | Bring out the flavour in foods, e.g. in processed foods such as crisps and flavoured rice |
| Flavourings | Thousands of different flavourings are used in food in small amounts; they need not be named individually in ingredient lists | Most processed foods | Restore flavour lost during processing or storage; add, or reinforce, flavour in foods |

| Glazing agents | Beeswax (E901) Carnauba wax (E903) | Sweets | Produce a shine on foods; may also have a preserving effect |
|---|---|---|---|
| Humectants | Glycerol (E422) | Cake icing and sweets | Prevent food drying out and becoming hard |
| | Sorbitol (E420) | Pastries and cakes | |
| Liquid freezants (processing aids) | Dicholorodifluoro-methane | Frozen strawberries | Help freeze food quickly |
| Preservatives | Nitrates and nitrites (E249-E252) | Bacon, ham, cured meats, corned beef, some cheeses | Prevent growth of micro-organisms which cause food decay and poisoning, extend shelf-life |
| | Potassium sorbate (E202) | Frozen pizza | |
| | Sulphur dioxide and the sulphites (E220-E227) | Dried fruit, dehydrated vegetables, fruit juices, sausages, fruit-based dairy desserts, cider, beer, wine | |
| Propellants | Nitrous oxide (also called dinitrogen monoxide) | Cream in aerosol cans | Help expel food from aerosol cans |
| Stabilisers | Carrageenan (E407) | Milk shakes, instant mousses, sausages, biscuits, pastries, meat pies | Prevent separation of emulsified oil and water; add smoothness and creaminess of texture |
| | Guar gum (E412) | Packet soups, coleslaw | |
| Sweeteners | Acesulfame K (E950) | Canned foods, yoghurts, soft drinks, table-top sweeteners | Sweeten foods |
| | Aspartame (E951) | Yoghurts, soft drinks, dessert mixes, table-top sweeteners | |
| | Isomalt (E953) | Sugar-free sweets | |
| | Mannitol (E421) | Sugar-free sweets, chewing gum | |
| | Saccharin (E954) | Soft drinks, pastries and cakes, sweets, ice-cream, reduced-sugar jams | |
| | Sorbitol (E420) | Sugar-free sweets, jam for diabetics | |
| | Thaumatin (E957) | Table-top sweeteners, yoghurt | |
| Thickeners and gelling agents | Gum arabic (E414) | Sweets | Thicken food and improve consistency |
| | Modified starch | Soups, sauces, pie-fillings | |
| | Pectin (E440) | Jam | |

them. The Scientific Committee for Food has no authority to carry out tests to verify the manufacturers' data.

## 'Use by and 'best before' dates

The dates marked on food labels are an important indication of whether a food is likely to be unfit or unpleasant to eat.

The 'use by' datemark is obligatory for highly perishable foods such as meat products or ready-prepared meals which could become a food safety risk. 'Use by' is intended as a clear instruction that the food should be eaten by the end of the date given and the shopkeeper may not legally sell food which has passed its 'use by' date. 'Use by' does not necessarily mean 'eat by' because cooking or freezing food before its 'use by' date will generally extend its life.

Foods which are unlikely to be a health hazard if they perish must carry an indication of how long they will last with a 'best before' date. This does not mean that when the date is reached the food may be dangerous; it means only that it is unlikely to be at its best.

'Use by' and 'best before' dates depend on both the shopkeeper and the consumer following the manufacturer's storage instructions. If they fail to, the food will spoil more quickly, and for foods which carry 'use by' dates the risk of food poisoning will be increased.

By law manufacturers must provide instructions about any special storage conditions required for the food, which are clearly worth following. Some retailers mark food with 'display until' dates, but these are instructions to shop staff; consumers should not confuse them with 'use by' and 'best before' dates.

## Information about processes used in the manufacture of foods

By law foods which have been dried, frozen, concentrated or smoked must be labelled as such (unless it is obvious that this is so). Foods which have been irradiated – a process (see pages 28-9) designed to kill harmful bacteria in foods and prolong their shelf-life – must carry the information that they have been 'irradiated' or 'treated with ionising radiation'. But generally it is difficult to discover how foods have been produced unless the manufacturer wants to tell you.

There is a growing controversy about whether foods which have been produced with the help of genetic engineering should be labelled accordingly. At the moment they do not have to be, although such products are reaching the supermarket shelves in increasing numbers. Already, certain brands of tomato paste are available which contain genetically modified tomatoes, and many vegetarian cheeses are now made with an enzyme which has been produced using gene technology.

## Nutrition labelling

Food manufacturers do not have to provide nutrition information on food packets unless they make a claim such as 'low fat' or 'high fibre', but they can supply such information voluntarily and most now do so. The nutrition information tells you how much fat, protein and carbohydrate there is in the food and also how many calories it contains. It may also give you information about other nutrients: sugar, starch, different types of fat, vitamins, minerals etc. You can use the table to choose foods that are 'healthier'. You can also use the information to check manufacturers' claims for their foods such as 'low fat', 'healthy', and so on.

If manufacturers provide nutrition information they must do so according to a specified format. There are two standard ones: a minimum list of the four key nutrients: energy, protein, carbohydrate and fat (the basic label); and a more detailed list of eight nutrients: energy, protein, carbohydrate and fat, plus sugar, saturated fat, fibre and sodium (the full label). Nutrient content must be stated per 100g (approximately 3½ oz) but manufacturers can also give information per serving (see table, page 138).

Different people find different elements of information on the nutrition information table useful. Slimmers may count calories, people who have been advised for medical reasons to follow a low-salt diet will need to check the sodium content. But the nutrition information table can be used by everyone endeavouring to adhere to a healthy diet.

## A guide to the nutrients in the nutrition information table

**Energy** The amount of energy which the food will give you when you eat it. Energy from food is measured in two ways:

● in calories (written as kcal on the label)

or

● in joules (written as kJ on the label).

One kcal (calorie) is roughly equal to four kJ.

**Protein** Protein is important for body growth and repair. Most people get more than enough protein for their needs, so you should not normally need to check the protein content of foods.

**Carbohydrate** This is almost entirely sugar and starch. Sugar and starch are different types of carbohydrate with different functions in the body.

**Starch** We should get most of the energy (calories) we need from starch instead of from sugar and fat. Foods with plenty of starch include bread, potatoes, breakfast cereals, rice and pasta. 'Starchy' foods used to be thought of as fattening, but weight for weight starch has less than half the calories of fat.

**Sugars** This means both the sugars which occur naturally in fruit and milk *and* added sugars. Added or refined sugars, such as those in table sugar, honey and fruit juice, can cause tooth decay if foods which contain a lot of them are eaten too often. The way you eat foods with a lot of added sugar is as important as how much of them you eat. Foods and drinks which are sipped, sucked and nibbled frequently throughout the day – such as soft drinks, sweets and biscuits – cause more damage because the sugar is often in contact with your teeth for a long time. Naturally-occurring sugars are not normally a problem for teeth. To find out whether sugar has been added to food, check the ingredients list (see page 127).

**Fat** This is the total amount of fat (and oil) in the food. A high intake of fat can lead to overweight and is the cause of many other health problems. Certain types of fat, such as the fat in oily fish such as sardines, are good for you. But general health advice is to cut back on fat. There are three main types of fat: saturates, monounsaturates and polyunsaturates. Most foods contain a mixture of each type but in different proportions. The full label tell you how much of the total fat is from saturates, while some labels also give a breakdown for monounsaturates and polyunsaturates.

**Saturates** This type of fat raises blood cholesterol levels and a raised blood cholesterol level increases your risk of heart disease. Watch out for saturates in meat products (pies, sausages, etc.), dairy products (butter, cheese, etc.), and cake and biscuits.

**Monounsaturates and polyunsaturates** Monounsaturates are neutral for heart disease and polyunsaturates lower blood cholesterol levels. It is better to eat food rich in monounsaturates (such as olive oil or rape seed oil) and polyunsaturates (sunflower oil and soya oil) than foods rich in saturates. But remember, they are still fats.

**Cholesterol** Cholesterol is a soft, waxy substance found in meat, dairy produce and eggs but not in foods which come from plants. In the early days of heart disease research it was thought that eating foods that contain a lot of cholesterol was the main cause of raised cholesterol levels in the blood, but nowadays it is recognised that it is the saturated fat in foods which is the main danger.

**Dietary fibre** Fibre used to be known as roughage. It helps prevent constipation, piles and bowel problems and may help to reduce blood cholesterol levels. Fibre comes from plants (there is no fibre in meat). Baked beans, kidney beans, high-fibre breakfast cereals, wholemeal bread and fruit and vegetables all contain a lot of fibre.

**Sodium** Most of the sodium in food is from salt (sodium chloride). Too much sodium causes high blood pressure. More than two-thirds comes from processed foods, so check the nutrition information table to cut back on it. The amount of sodium in foods is normally much smaller than the amounts of other nutrients, so the figures on the nutrition label are often less than 1g. But these small amounts matter.

**Vitamins and minerals** Vitamins and minerals are vitally important for health, but you need only small amounts and most people get more than they need. Vitamins and minerals have to be shown on the label as a proportion of a Recommended Daily Allowance (RDA).

Amounts of vitamins and minerals can only be shown if the food contains more than 15 per cent of the RDA per 100g.

## What to look for in the nutrition information table

You do not need to look at all the numbers in the nutrition information table to make healthy choices. Some nutrients are more important than others. First, look at the amount of fat and saturates in the food. Choose foods with the least fat, particularly the least saturates. Next check the sodium content: the less sodium, the better. Then check the fibre and lastly the sugars. There is no need to pay much attention to calories unless you are trying to lose weight. Eating healthily does not mean avoiding calories.

Everyone over the age of five can check the nutrition information in this way: fat (and saturates), sodium, fibre and sugars. Children, teenagers and the elderly are particularly prone to tooth decay and therefore need to take special care to avoid sugar. For infants and young children under two, sugar and sodium are the main nutrients

**Nutrition: basic label**

| Nutrition information | |
|---|---|
| | Typical values per 100g |
| energy | 1889 kJ<br>450 kcal |
| protein | 3.3g |
| carbohydrate | 70.4g |
| fat | 17.2g |

**Nutrition: full label**

| Nutrition information | | |
|---|---|---|
| | Typical values | |
| | per 100g | per pack 35g |
| energy | 1889 kJ<br>450 kcal | 661 kJ<br>158 kcal |
| protein | 3.3g | 1.2g |
| carbohydrate<br>of which<br>sugars | 70.4g<br><br>1.1g | 24.6g<br><br>0.4g |
| fat<br>of which<br>saturates | 17.2g<br><br>6.2g | 6.0g<br><br>2.2g |
| fibre | 2.7g | 0.9g |
| sodium | 1.5g | 0.5g |

to check out – the less the better. The amount of fat and fibre in foods is not so important because children of this age have small stomachs and there is a risk that a low-fat/high-fibre diet might fill them up without providing enough calories. From two years on children should gradually move towards a low-fat/high-fibre diet.

For some people it may be worth looking at the mineral and vitamin content in the nutrition information panel to make sure they are getting enough. In particular, pregnant women and women planning on getting pregnant should try to ensure that they get at least 300µg per day of folic acid, i.e. 50 per cent more than their normal requirement. This is because a lack of folic acid at the time of conception and in the early stages of pregnancy has been shown to be associated with neural tube defects (such as spina bifida) in babies.

## How much is a lot or a little?

In general the less fat (particularly saturates), less sodium (salt) and the more fibre, the healthier your diet, but the table of guideline intakes below gives you an idea of a healthy amount to eat, on average, each day.

**Guidelines for daily fat intake**

|  | Men | Women |
|---|---|---|
| Fat | 95g | 70g |
| Saturates | 30g | 20g |
| Sodium | 2.5g | 2g |
| Fibre | 20g | 16g |
| Sugar | 70g | 50g |

These guideline intakes, based on government recommendations, are for an average-sized man and woman involved in an average amount of physical activity. In real life, everyone is different and has different nutritional needs. A tall woman may need more than a short man. Someone who is very active may need more than someone who does not take much exercise. So the figures are only a very general guide.

You can use these guideline intakes to work out how a food fits into your whole diet. For example, to work out how much of your daily fibre needs you will get from a bowl of breakfast cereal, compare the amount of fibre you get in a serving with the guideline intake. Alternatively, here are some rules of thumb to give you an idea of how much is 'a lot' or 'a little' of the most important nutrients for making healthy choices:

| 'A LOT'<br>These amounts or more: | 'A LITTLE'<br>These amounts or less: |
|---|---|
| 10g of sugars | 2g of sugars |
| 20g of fat | 3g of fat |
| 5g of saturates | 1g of saturates |
| 3g of fibre | 0.5g of fibre |
| 0.5g of sodium | 0.1g of sodium |

When using these rules of thumb you should look at the amount you get in a serving of the food. However, for snacks and foods such as margarine which you eat in relatively small amounts it is better to look at the amount of nutrient you get in 100g.

### How to use the nutrition information table to compare foods

You can also use the nutrition information table to compare foods. To choose between foods you eat in different amounts, compare the amount of nutrient you get in a serving of each food. For example, when choosing between two ready-meals look to see which has the most fat.

For foods which you are likely to eat in roughly the same quantity compare the amount you get in 100g of the food. For example, when choosing between two breakfast cereals look first at the amount of fat per 100g (it is likely that they will have roughly the same); then look at the amount of sodium per 100g they both have: it is likely to be very different.

## Advertising claims

Look at most food packets and you will see that the majority of the space is taken up by advertising claims. Most packets, unless they are made of clear plastic, show a picture of the food – usually looking highly appetising. There will probably be claims in addition about how good it tastes and what wholesome ingredients it contains.

The packaging is likely to carry other types of claim as well. The number and variety of health-related claims for foods has increased dramatically in recent years. They range from the general, such as 'healthy' or 'diet', to specific, such as 'low-fat' or 'helps lower cholesterol'. But you will also find claims about how the food has been produced or even traded. It has long been possible to buy eggs which claim to be 'free-range', but nowadays many other animal products claim to have been produced according to humane farming methods rather than under intensive factory régimes.

As concern grows about the impact of food production on the environment and of modern food production methods on food, farmers and manufacturers have been under pressure to respond. Many foods are now labelled 'organic' or 'natural'. Almost all canned tuna now claims to be 'dolphin-friendly'.

Recently, foods have started to carry claims that they have been 'ethically traded' so that Third World producers get more for their produce than usual and hence should not be among those that get away with rock-bottom wages or employing children.

The health-related claims on food packets can be divided into those which are merely about the nutrient content of the food (nutrition claims) and those which suggest that eating the food is good for your health in some way (health claims).

## Nutrition claims

It is best to treat nutrition claims such as 'low fat', 'reduced sodium' and 'high-fibre' with caution: some of them are little more than marketing hype. Many are ambiguous and some are frankly misleading. They can often be highly selective, too, highlighting the good points about a food whilst failing to mention less desirable features.

The food labelling law covering the use of such nutrition claims is seriously out of date. While there are specific rules which cover the use of some claims, such as 'low in calories', 'low in cholesterol' and 'high in vitamins', these rules need changing in the light of new nutritional research. For example, 'low in cholesterol' claims can be made even when the product is high in saturated fats, which makes little nutritional sense (see pages 136-7).

Moreover, there are no specific rules for fat and sugar claims – two of the most common types of claim nowadays. Nor are there rules for claims about salt and fibre, which, after fat, are the most important nutrients to watch for if your aim is to eat healthily.

Labelling law is now the responsibility of the European Union, which has been drawing up, scrapping and rewriting proposals for a directive on food claims since 1981. Meanwhile the UK government could have modernised its own rules, at least in respect of foods manufactured in Britain, but has failed to do so.

The main problem with most nutrition claims is that they are ambiguous. When a packet of sausages claims to be 'low-fat', more likely than not they will still have a lot of fat in them. They will just have a little less fat than other sausages. Similarly most 'low-fat' spreads still contain 25-40 per cent fat. But some low-fat products really do have only a little fat in them. 'Low-fat' yoghurts on the whole really do have less than 3g of fat per serving. The rules of thumb in the table above are a good way of gauging how much of a nutrient a food really contains.

Some nutrition claims are even more deliberately ambiguous. If you see 'light' or 'lite' on a packet you might assume that its contents are lower in calories or perhaps fat. In some cases you would be right, but in other cases the manufacturer seems to mean that the product is light in colour. There are no rules about using a 'light' or 'lite' claim.

The other major problem with nutrition claims is that they are selective. Many 'high-fibre' breakfast cereals contain a lot of salt. A single bowl of bran flakes will give you about 0.3g of sodium, which is about an eighth of your daily guideline intake. All you can do is check the nutrition information table.

## Health claims

Like nutrition claims, health claims on food packets are becoming increasingly common. They range from the incredibly vague, such as 'good for you', 'nutritious' or just plain 'healthy', to the more specific – such as 'helps lower cholesterol', 'good for your digestion' and so on.

As for nutrition claims the law on health claims is seriously out of date. The only really significant restriction on the use of health claims is that manufacturers are not supposed to be allowed to say that a food can prevent, treat or cure a disease. So, for example, a food cannot claim to prevent heart disease. But manufacturers increasingly suggest that their foods help prevent the disease without actually saying so in so many words. Heart-shaped bowls for breakfast cereals, detailed explanations of how particular nutrients are 'good for your heart' and so forth are increasingly common. While references to heart disease seems to be the most popular at the moment, other diseases – such as neural tube defects, osteoporosis and even cancer – have also begun to be mentioned.

In the US the government has recently bowed to increasing pressure from food manufacturers to allow them to use more explicit health claims and has passed a law which allows health claims to be made, under strict conditions, in seven areas:

(1) calcium and osteoporosis
(2) fat and cancer
(3) saturated fat and cholesterol and heart disease
(4) fibre and cancer
(5) fibre and heart disease
(6) sodium and high blood pressure
(7) fruit and vegetables and cancer.

This is a reasonably good guide to where there is enough scientific evidence for a health claim to be made about a food. So, for example, if a food claims to contain a natural ingredient – a bacterial culture – which can 'actively reduce blood cholesterol levels, which are a risk factor for heart disease', it is unlikely that there is sufficient evidence to support this claim.

## Other types of claim

In general the law on claims is weak and feeble. Only in the case of 'organic' claims does the situation seem to have been reasonably well managed. There are now strict European rules for when a food can and cannot claim to be organic. To counter the proliferation of claims which consumers have difficulty in evaluating by themselves, some voluntary organisations have introduced endorsement schemes for foods. One of the best-known of these schemes is the Vegetarian Society's V logo for vegetarian foods. The RSPCA has recently introduced a 'Freedom Food' logo for foods from animals 'reared, handled, transported and slaughtered compassionately'. Such schemes ought to be helpful to consumers looking for particular types of foods but in some cases the organisation running the schemes is closely linked to the manufacturers whose products they endorse.

The best advice to consumers remains, as for all labelling issues, *caveat emptor* (buyer, beware!).

# CHAPTER **4**

# Serious food complaints

THE FOOD industry is regulated by laws which cover the food chain from the farm to the shop. We all buy food, and these laws protect us whether we are buying the week's groceries from the supermarket or simply a packet of crisps, a bar of chocolate or a pint of milk. But what if the food makes us ill, or is wrongly priced, labelled or weighed, or we are simply shortchanged?

Protection comes in two forms:

(1) **civil law** is the branch of the legal system which is of most use to the individual seeking redress in the form of compensation. The Sale of Goods Act 1979 (as amended) and Part I of the Consumer Protection Act 1987 are two examples of Acts of Parliament which offer civil protection. The civil law is concerned with rights and duties that relate to individuals in their dealings with other individuals (including companies and other groups of people). If you suffer loss because someone else breaches civil laws then you have a right to redress and are entitled to take that person to court if need be. The main areas of civil law are **tort** (which includes negligence) and **contract**. The courts which deal with civil claims are the **county court** (which includes the **small claims court**) and the **High Court**.

(2) **criminal law** is the branch of the law which is concerned with offences against the public, such as the Food Safety Act 1990, the Trade Descriptions Act 1968 and part II of the Consumer Protection Act 1987. Criminal law is generally enforced by the police, but the specific criminal law affecting consumers is enforced by public

authorities such as Trading Standards Departments and Environmental Health Officers (both based at your local council offices). You cannot usually get compensation directly by reporting a criminal offence such as a false trade description, but evidence of such wrongdoing will lend weight to your complaint and sometimes a compensation order may be made as part of the prosecution. The courts which deal with criminal matters are the **magistrates' courts** and the **crown courts**.

## Legal protection

A number of Acts of Parliament set out the rights and duties of specified people in specified circumstances. The five major statutes which protect you as a buyer and consumer of food are:

(1) **Sale of Goods Act 1979** (as amended by the **Sale and Supply of Goods Act 1994**) This offers protection to buyers of all types of goods, including food, and is perhaps the most important single piece of consumer legislation. It sets out the obligations to businesses that sell goods to consumers. Whether the contract of sale is written or simply spoken, the Sale of Goods Act 1979 implies terms into that contract. It says that when you buy goods from a business those goods must:

- fit any **description** given of them
- be of **satisfactory quality**
- be reasonably **fit for their purpose**
- correspond with any **sample** that the customer was shown.

(2) **Consumer Protection Act 1987** and, in Northern Ireland, the **Consumer Protection (N.I.) Order 1987** These state that manufacturers, which include food producers, are strictly liable if the products they make are defective and cause personal injury, or damage to property over £275, and permit the purchaser to claim compensation in the event of any such occurrence. So, if your apple pie contains a piece of glass which injures you, or your ready-prepared meal catches fire and damages your oven and your kitchen, you may have a claim. The crucial provisions for getting compensation are that:

- you have to prove that the food product was **defective** and that it **caused** the injury or damage that you are complaining about

- a product is considered **defective** if it is **less safe** than consumers generally are entitled to expect, so to be caught by this food must be unsafe, not merely of poor quality. Food could therefore be defective under this Act if the instructions for cooking or any warnings are inadequate, making it unsafe to eat
- it does not take away the existing liability of retailers under the Sale of Goods Act 1979
- the term **product** includes food, but not game and primary agricultural produce. To be covered food must have gone through some industrial process which changes the nature of the food. Simply harvesting tomatoes would not constitute a process. Hence, a farmer is not liable for selling rotten peas, but a food manufacturer would be liable for producing defective pea soup or defective frozen peas
- you cannot claim for damages to the product itself. So if the packaging of a ready-prepared meal melts into the food in your oven but does not damage you, the oven itself or indeed anything else, say, this law will not help you, although you may have a claim against the retailer for a refund under the contract of sale (Sale of Goods Act 1979).

The Consumer Protection Act also makes it a criminal offence for traders to sell anything other than safe goods, and to give price indications and make bargain offers which are misleading.

(3) **Food Safety Act 1990** This Act and regulations under it provide the criminal law which ensures food standards and hygiene wherever food is manufactured, prepared or sold, as well as other aspects of food and drink handling. It makes it a criminal offence to sell food that is not fit for human consumption or which is not of the substance or quality demanded by the consumer and covers false and misleading labels, descriptions and advertisements. Both Trading Standards and Environmental Health Officers (TSOs, EHOs) police this area.

(4) **Trade Descriptions Act 1968** This makes it a criminal offence for traders to make false statements about the items they sell. This covers many issues concerning the physical characteristics of foods and their origin – for example, statements about quantity, size and virtues (such as health claims), as well as statements about who the

food was produced by and where (such as Scottish smoked salmon, or Welsh lamb). Report any problems to the local Trading Standards Department.

(5) **Weights and Measures Act 1985** and, in Northern Ireland, the **Weights and Measures (N.I.) Order 1988** This stipulates that (1) all weighing and measuring equipment should be accurate; (2) many commodities, especially foods, should be marked with the quantity supplied, and that certain commodities should be sold only in certain specified quantities; and (3) all aspects of quantity control in trade should be open to official inspection. Again, Trading Standards are responsible for prosecutions under this legislation.

Some other laws are worthy of note:

- **Unfair Contract Terms Act 1977** This says that any attempt to take away or restrict your Sale of Goods Act rights is ineffective. And the Consumer Transactions (Restrictions on Statements) Order 1976 makes it a criminal offence for anyone to include a term in a contract or to put up a notice attempting to take away or restrict those rights.

- **Supply of Goods and Services Act 1982** This Act (or the equivalent common law in Scotland) defines the standard of the service consumers can legally expect when engaging a trader to provide a service; it also cover the quality of the materials used and other criteria such as price and timescale. So, if you employ caterers in your home for a party, say, this Act will protect you. The service must be performed with reasonable skill and care (this includes the food preparation) and all the food supplied must be of satisfactory quality and fit for its purpose.

- **Consumer Protection (Cancellation of Contracts Concluded Away from Business Premises) Regulations 1987** This gives purchasers a seven-day cooling-off period during which they have the right to cancel certain contracts made during an unrequested visit by a salesman to their home. This does not cover regular deliveries by roundsmen such as your milk or bread delivery but would include a salesman who turns up unannounced.

# Buying food

## Consumers' rights

Every day we make contracts without putting them in writing, or even speaking a single word: we buy food in shops, pay cab or bus fares, and so on. All these have the same standing as written contracts and are governed by the same laws as other contracts. Whether written or verbal, a contract is an agreement that can be enforced by the law and gives rise to rights and responsibilities for those involved.

Some rights are always implied into certain transactions: for example, when you buy goods, including food, or when you employ a trader to do work for you, the rights embodied in the relevant Acts of Parliament automatically come into play and many can *never* be taken away.

Spoken agreements can cause problems if it later becomes necessary to prove the precise terms agreed, such as the price or a delivery date. It is always worth keeping receipts for food in case you need to prove later where and when it was bought and how much you paid.

The Sale of Goods Act 1979 imposes a strict liability on retailers. This means that the food they sell must satisfy each of the various implied terms below. If it does not they will be liable for breach of contract. A retailer is not allowed to claim that he did not know about a defect and could not reasonably be expected to have discovered the problem. So although the problem may have been caused by someone else back along the food chain, and may be something which the retailer could not possibly have noticed, the shop is still fully responsible because of his contract with you the customer.

- Food must be **as described**: for example, 'fresh' turkey must not have been frozen. This is not only a breach of your rights but can create other problems: a fresh turkey, after all, can cost about twice the price of a frozen one – but it could also be a health hazard. It can be dangerous to eat a frozen turkey which has not had time to defrost properly before cooking, or to try to refreeze it.

- Food must be of **satisfactory quality**: 'fresh' food must actually be fresh and not stale, food should last a reasonable length of time taking into account what it is, certainly up to and including its 'use by' date if it has one, and it should be free from any contamination.

149

If food is stale – when you get your shopping home and find, for example, that the bread is stale, or the vegetables are mouldy – you are entitled to complain, but do act quickly. If food does not last as long as it should – you open your milk, say, only to discover that it has gone off before the 'use by' date – you are entitled to complain, as long as you have stored it correctly. And if food is contaminated in some way – you open a bag of flour to find it infested with weevils, or you find a piece of string, or worse, in a can of baked beans, or you find any kind of foreign body in your food – you can complain (see below under 'Foreign bodies').

● Food must be **fit for its purpose**: hence, fit to eat and be capable of being used for any purpose specifically promised.

● Food must correspond with any **sample** you were shown before buying: so if you order a cake from a shop having seen a sample, the one you get must be made in the same way and be of the same size, colour and shape.

## Food deliveries

There is no legal requirement that food be delivered within a fixed period, so it is up to you to tell the seller if you have a deadline. The Sale of Goods Act 1979 says that the seller must deliver the goods to you within a 'reasonable' time. There are no hard and fast rules about what is 'reasonable'. It depends on the circumstances – the type of food, its availability and so on. There are laws governing the way food is stored and transported including temperature, handling etc. (see 'Food safety', below). Note that:

● if the date of delivery is important you should make 'time of the essence' in writing when making the contract. Then, if delivery is delayed by even a day, you are entitled to cancel the contract and receive a full refund of the price paid. And if it costs you more than the price you originally paid or agreed to get the same food elsewhere, you can claim the extra cost from the original supplier

● if no specific date was agreed in this way you can give notice in writing to the retailer imposing an ultimate time limit. The period specified in writing must be reasonable, and if the food does not arrive by that deadline you may cancel the contract and get your money and extra costs back. You would then be within your rights to go to another supplier who has the food in stock and get it immediately. If in doing so you have to pay more, the additional cost can be recovered from the first retailer

- sometimes it is clear from the nature of the food ordered (a Christmas turkey, or an Easter egg, say) that it is required by a particular date. So even if you have not specified a date, if such food does not arrive in time then you are entitled to cancel the contract and get a refund of your money plus any extra costs, unless of course you decide you still want the food.

## Mail order

The number of companies offering food by mail order is increasing. If you have any quibbles about quality, or if you become ill after eating some food bought by mail order, say, you have exactly the same rights as you have when you buy food from a shop, so you should complain in the same way as if you had bought the food in a shop. The mail order company is the retailer, therefore:

- you are fully covered by the Food Safety Act 1990, which provides protection against contamination, sets out rules on temperature control and so on, in the same way as food sold in shops (see 'Food safety', below)
- you are fully covered by the Sale of Goods Act 1979, so the food must be of satisfactory quality and be reasonably fit for its purpose, and should correspond with its description in advertisements or catalogues
- you are fully covered by the Consumer Protection Act 1987, so if something you have bought turns out to be dangerous and causes illness or injury you can claim against the manufacturer or the company which produced or processed the food (see page 146).

The Food Safety (Temperature Control) Regulations 1995, made under the Food Safety Act 1990, impose a general requirement that nobody shall keep foods which are likely to support bacteria or toxins at more than 8°C. There is an exception to this where food is sent by mail order. However, the Regulations prohibit anybody from supplying food by mail order by post or other carrier at a temperature which is likely to give rise to a health risk.

The Department of Health has also issued guidance on these Regulations, which recognise that a limited amount of time outside a controlled chilled environment would not have adverse consequences for human health. Some foods are stable, even for long periods, unchilled. So mail order companies can only send the sort of food that is suitable for sending by post and it must be packed and handled in such a way that meets these general requirements. If you think food you have received by mail is unfit for consumption, contact your Environmental Health Department. If the quality has suffered, contact Trading Standards.

There is no legal requirement that food sent through the mail must arrive within a particular time (see under 'Food deliveries', above). The law states only that it must arrive within a 'reasonable' time, which will depend on the type of food. The British Code of Advertising Practice (BCAP) states that mail-order goods should arrive within 28 days of receipt of order. When you order food in this way:

- if you need it by a particular date, make this date 'of the essence' when you send your order
- keep a note of the date (as well as the company's name and address, and what you ordered) so that you will know when the date you have mentioned arrives or the 28 days are up
- if the food fails to arrive by the required date, or within the 28 days set out in the BCAP, you can cancel the contract and recover your money
- if the trader is a member of a trade association, contact it for help in sorting out your problem. If it is not a member of any particular association, contact the Advertising Standards Association (ASA). The ASA administers the BCAP, and although this does not have the force of law it is a good indicator of what length of time is 'reasonable'.

If food arrives damaged, you will normally find that mail order companies which belong to trade associations agree to replace the damaged food free of charge. But there may be no protection if the company is not a member of a trade association and therefore does not subscribe to a code of practice. In such circumstances your rights will depend on whether the supplier can prove that the food left its premises in perfect condition.

However, the Sale of Goods Act 1979 says that once goods have been handed over to a carrier (this may be the Post Office or a delivery company) they become the property of the buyer, who must bear the risk of accidental loss or damage in transit. This may seem harsh, but the Act also says that the supplier should make sure those goods are adequately covered by insurance during transit (by sending them by registered post, say). So, if it is clear that the food was damaged in transit, you may have a claim against the carrier. In the case of the Post Office, you should complete form P58 (available at any post office counter) and send it to the Head Postmaster. Remember to keep a copy of what you write. If you can show that the parcel was damaged in the post, you are entitled to compensation on a sliding scale, but it may not cover the actual cost of the food. If the food arrives damaged and the supplier has not taken out any insurance protection you can refuse to accept the goods and demand your money back.

## Taking action

If the food you bought has made you ill you must act quickly and get together evidence to show that it was the food that caused the illness. If the food simply is not up to scratch you are entitled to financial compensation because the seller has broken a condition of the Sale of Goods Act 1979. This compensation should be in the form of a full refund of the price paid. With complaints about food it is extremely important that you act quickly as you can lose this right if you leave your complaint too long.

For example, if you buy some fresh meat and immediately put it in the freezer when you get home, and three months later when you come to defrost some of it you discover that it is not what you asked for or has gone off, you would find it difficult to claim anything unless you can prove the meat was unsatisfactory when you bought it and that the cause of the problem was nothing to do with you.

If the retailer does offer a replacement or credit note you may prefer to accept this, particularly if it is too late to get a refund – the decision is entirely yours – but it cannot force you.

The usual way to reject goods is to return them to the shop and demand an immediate refund. If the shop refuses, send a letter making it clear that you are rejecting the food and seeking a refund of the price paid.

## Food labelling

Consumers are meant to be protected by the various laws which govern food content and labelling. But gaps in the law or vagueness about how it should be interpreted mean that, even if foods are labelled correctly, there is still a chance that you will be misled.

Food producers may want to make expensive ingredients go further by mixing them or swapping them with cheaper ones and not making it obvious to the customer. Other foods can mislead by the way they are labelled.

The **Food Safety Act**, the **Trade Descriptions Act**, the **Consumer Protection Act** and the **Weights and Measures Act** all aim to protect consumers by making it an offence to describe food in a way which could mislead. Food labelling regulations (see page 156) also control what appears on food labels.

Breaches of the regulations occur quite frequently. For example, Durham Trading Standards found that in ten out of the 11 pizzas it bought from pizzerias and two out of the six sandwiches bought from garage shops, 19-25 per cent of the so-called 'cheese' was really cheese substitute – made from skimmed milk and vegetable oil. Several prosecutions resulted. In Sheffield, an analysis of 54 samples of so-called minced beef found that 13 in fact contained other meats such as pork and lamb. The same public analyst found pork in minced meat being sold as halal. The West of England Food and Quality Standards Group found 114 out of 221 sandwiches wrongly labelled – for example, describing a sandwich as 'ham' when it contained pork shoulder, and another as 'beef' when it contained reformed meat.

The food labelling laws, covering ingredients, nutrition, 'use by' and 'best before' dates etc., all of which are explained in Chapter 3, are enforced by TSOs.

The law requires most food labels to carry a datemark. This can be a 'best before' or a 'use by' date. A few foods do not have to carry a datemark.

- It is not illegal to sell food simply because its 'best before' datemark has expired. The food could still be sound and wholesome, but it may simply be past its best. This mark merely alerts you to this possibility.

- Food with 'use by' dates must be withdrawn from sale once that date has passed. If you are worried about a particular datemark, tell the shopkeeper or manager, and the TSOs or the EHOs at your local council. It is an offence to sell foods bearing an expired 'use by' date and for anyone to change it.

- Whether the 'use by' date has expired or not it is an offence under the Food Safety Act 1990 to sell any food that is not fit for human consumption, or not of the nature, substance or quality that a purchaser might expect from the labelling or packaging.

- If the food is not in the condition expected, or if it has gone off and is not fit to eat, then you also have the right to claim under the Sale of Goods Act 1979 (see 'Legal protection', above).

## Whose fault?

Recent years have seen a staggering increase in food poisoning cases, but these should not be blamed totally on consumers. The food industry believes that there are areas where practices in the home could be improved, but often the problems lie with mistakes made earlier in the chain. Cross-contamination during production is thought to be a frequent cause of food poisoning.

Labels must give instructions to ensure that the food remains in good condition and how to make the best use of it. If you do not follow the instructions and the food goes off, you may be unable to complain effectively. So, if a label says 'refrigerate after opening' or 'keep in a cool, dark place' and you do not follow this advice, it could be your fault.

The safest rule is that any complaint you have about a food should be made first to the shop where you bought it or, if the food was pre-packed, to the address on the label. If this does not satisfy you, or if you think there is something more serious that could affect the wider public, contact your local council. In England and Wales the Trading Standards Department of the county council enforces food labelling laws. In metropolitan areas it is the district council; in London it is the

London boroughs; and in Scotland and Northern Ireland it is the Environmental Health Department of the district council. You will find these listed in the phone book.

## General requirements for food labelling

The Food Labelling Regulations 1984:

- prohibit the misleading presentation of food;
- require all food to be marked or labelled with the name of the food, a list of ingredients, an indication of minimum durability, any storage conditions or conditions of use, the name and address of the manufacturer or packer or of the seller, the particulars of the place of origin of the food (if failure to state this could mislead the purchaser to a material degree as to the origin of the food), and instructions for use, if it would be difficult to use the food without these;
- require, where the labelling of a food places special emphasis on the presence of a low content of an ingredient in a food, an indication of the minimum or maximum percentage of that ingredient in the food;
- make special provisions for the labelling of food which is not pre-packed and similar foods, fancy confectionery products, food which is packed in small packages and food for immediate consumption;
- specify additional labelling requirements for alcoholic drinks and for food sold from vending machines;
- specify requirements as to the manner of marking or labelling food;
- prohibit certain claims, and impose conditions for making other claims, which include claims relating to foods for particular nutritional uses, diabetic claims, slimming claims and medicinal claims;
- impose restrictions on the use of certain words and descriptions in the labelling or advertising of food.

What's in a name? Just because an ingredient features in the main brand name of a product, it does not mean that the product contains a lot of that particular ingredient. A food can contain larger amounts of other things which the manufacturer need not mention in the name. Also, ingredients (except for additives) which are themselves part of an ingredient which makes up less than a quarter of the end product do not need to be listed. For non-pre-packed foods in restaurants or from takeaways, there is of course no requirement for any ingredient at all to be listed. All of this makes it impossible for you to know for certain what is in the foods you eat.

## Weights and measures

The law requires that food is sold in accordance with the advertised quantity and weight. The Weights and Measures Act 1985 and specific orders made under it tackle a variety of issues. All of these are policed by TSOs. For example, the Weights and Measures (Miscellaneous Foods) Order 1988 stipulates that foods which are pre-packed must be labelled with prescribed quantities and the containers marked with the quantity; if the food is not pre-packed, it must be sold in specific quantities. Under the Price Marking Order 1991 non-pre-packed foods (for example, vegetables sold loose) are required to be marked with a unit price in either kg or lb, 100g or ¼lb.

'Pre-packed' means 'made up in advance ready for retail sale in or on a container'. The list of such foods is long, ranging from rice to biscuits, breakfast cereals and jam. It states the prescribed quantities in which the foods must be packed, exceptions to this, and exemptions from quantity marking for both pre-packed and non-pre-packed foods.

Regulations on units of measurement have been implemented following an EU Directive which imposed a general ban on the use of imperial units of measurement. These have been prohibited since 30 September 1995 with a few exceptions and temporary arrangements. For example, until 1 January 2000 the pound and ounce may still be used for selling loose goods from bulk, but after that date metric measurements must be used.

## Foreign bodies

If you buy food that is contaminated you should take it to the EHO at your local council, who will investigate the matter and may recommend a prosecution. Whatever the EHO does, you are still entitled to claim compensation for any injury or for the stress caused. This gives rise to two possible claims (see 'Legal protection', above):

- under the Sale of Goods Act 1979 against the retailer. If food is not of satisfactory quality or fit for its purpose the retailer is in breach of contract. Claims such as this, even if there is no physical injury, can sometimes lead to substantial compensation for the stress and trauma incurred.
- under the Consumer Protection Act 1987 against the manufacturer, producer or importer. As long as the food has been through some form of 'process' it must be as safe as you are entitled to expect and there is no need for you to prove that the manufacturer was at fault in any way.

You should complain in writing straight away, and make sure you get the necessary evidence to show that the food was contaminated. If you have been injured you may need a report from your doctor to back up your claim.

## Food safety

### Illness

Hygiene regulations govern the food chain from slaughterhouses through processing plants and manufacturers to retailers. The consumer is the last link in the chain. But somewhere along that chain things can and do go wrong and illness can result.

Most of the meat and other food hygiene rules make it a criminal offence if a manufacturer or retailer fails to comply and it is the TSOs and EHOs at the local council who police these rules. They have the authority to inspect and take samples of food at any stage of the production and selling process. Local authorities collect a wide range of food samples for testing by public analysts to try to prevent food fraud, and they can prosecute producers and retailers who deliberately mislead consumers. However, the processes are far from straightforward.

If you think food you have eaten has made you ill you should tell your doctor immediately. Although it can be difficult to pinpoint the cause of illness, since symptoms can take up to three days to appear, the fact that you may be ill after a particular meal will make it more obvious that the food was substandard.

You can claim compensation for your pain and suffering, and any loss of earnings and other expenses you incur as a result, including a refund of the cost of the meal. Tell your doctor and get a report on the cause of your illness.

The Food Safety Act 1990 makes it a criminal offence for a retailer to sell food which is unfit for human consumption. You should tell the EHO at the council offices (in the phone book). EHOs can investigate the incident and the retailer could be fined and forced to compensate you. The EHO can even close down the business if it is found to be responsible.

The Act applies to major retailers and small shops, as well as restaurants, cafés and food sales at charity bazaars and other fund-raising events, and right back along the food chain to the producer. Enforcement is carried out by the local authorities, so always contact the local council with your complaint as soon as possible.

The main offences under the Food Safety Act are:

- selling, or possessing for sale, food which does not comply with 'food safety requirements', rendering food injurious to health;
- selling, to the purchaser's prejudice, food which is not of the nature, substance or quality demanded – for example, if pink trout is sold as 'salmon', or sheep's-milk yoghurt is actually made from cow's milk, or beef stew has too little beef in it, and so on;
- falsely or misleadingly describing or presenting food – this covers statements or pictures which are untrue or which may be true but are misleading. There are also detailed food labelling regulations (Food Labelling Regulations 1984) (see 'Food labelling', below and Chapter 3).

The 'food safety requirements' are that all food, not just sold in shops but all the way through the food chain, must not:

- have been rendered injurious to health. This means that it is of a kind that would harm a substantial part of the population. Four people were killed in the UK in 1995 by allergic reactions to nuts in food bought in restaurants and takeaways. But if you happened to be intolerant to a specific ingredient, such as certain kinds of nut, but the general population is not adversely affected, this would not be covered. However, there may be a separate breach of food labelling law for failing to list the ingredients, although this requirement would not help if the nuts were part of another ingredient which makes up less than 25 per cent of the final product, such as the almonds that make up marzipan (see Chapter 3). It is also an offence intentionally to do anything which would make food harmful by adding something or removing something.

- be unfit (for example, it was putrid or toxic).
- be so contaminated, whether by extraneous matter or otherwise, that it would be unreasonable to expect it to be eaten (for example, it contained a rusty nail or antibiotic residues).

The Food Safety (General Food Hygiene) Regulations 1995 require that the owner of any food business ensures that food is handled or maintained hygienically at all stages. This obligation covers temperature for storage and display, cleanliness and prevention of contamination and breach of the Regulations is a criminal offence. And any business which sells unsafe food may have to pay compensation to consumers who have been injured by that food under the Sale of Goods Act 1979 (as amended), or under the Consumer Protection Act 1987 (this covers manufactured food but not unprocessed agricultural produce); similarly, the criminal courts can make an order requiring a convicted offender to pay compensation to a person who has suffered loss or injury as a result of the criminal act.

# Shopping

## Prices displayed

If you see food on display in a window or in the shop you cannot insist on paying the marked price. Under the civil law a price displayed on the shelf, on the goods themselves, in an advertisement or on a price list, or in the window of a shop is called an invitation to treat, that is, an invitation for the public to go to the till and make an offer to buy the goods at that price. But if the shop realises it has mispriced the goods displayed, you cannot insist on buying them at the price marked:

- in fact, a trader can refuse to sell anything to you for no reason at all
- you can try offering the market price, and the trader may agree to sell the item to you at this price
- you can always bargain by offering any price for goods, and if the shop accepts your offer then a contract is made
- if the shop does sell you food at the wrong price by mistake, it cannot insist you pay the extra later.

Under the criminal law the trader offering the goods at the 'wrong' price (where the 'real' price that he will charge is higher) could be liable to prosecution. The Trade Descriptions Act 1968 makes it an offence to give an indication that goods exposed for sale are being offered at a lower price than that at which they are in fact being offered. Also, the Consumer Protection Act 1987 makes it an offence to give a 'misleading price indication'. You should report cases such as this to the Trading Standards Department at your local council offices. But if the price displayed was a genuine mistake then the shop would not be prosecuted.

## Sales and promotions

If a shop misleads its customers into thinking that they are getting a better deal than they really are the shop may be committing a criminal offence. The Consumer Protection Act 1987 sets out guidelines on misleading prices. For example, you may see the price of a large pack of smoked salmon cut from the previous price of £20 to £10. If any shop is making a comparison with its previous prices:

- the previous price should be the last at which the food was available in the previous six months
- the food should have been available at the higher price for at least 28 consecutive days during the last six months
- the food should have been on sale at that previous, higher price for that period at the same shop.

The law aims to prevent customers being misled. However, retailers can give notice that these conditions do not apply, as long as they spell it out clearly – for example, that the earlier price was only available for one week previously, or that that price had been available only in selected branches. If they do not give you this information and you suspect that you have been misled, report the matter to the Trading Standards Department for the area where the shop is.

## Health claims

It is illegal for claims to be made that certain foods can prevent or cure illness or disease. If they did they would count as medicines, and would therefore need to meet the different (and very stringent) criteria which would allow them to be licensed as medicines.

You may think you can rely on nutritional claims such as 'low fat' or 'reduced sugar' as a quick way to identify healthier foods. But arcane laws and inconsistent labelling make it difficult to choose which foods have the healthiest ingredients.

Manufacturers that wish to make a nutritional claim for a food product have voluntary guidelines to follow:

- all figures quoted should be per 100g or 100ml (for example, for the amount of fat or sugar that the food contains)
- a 'reduced-fat' product should contain no more than three-quarters of the amount of fat in the regular product
- a 'low-fat' product should not contain more than 5g of fat per 100g and per normal serving
- if a product claims to have 'reduced saturates', levels should be no more than three-quarters of those in the regular product
- if a product is 'low in saturates', there should be no more than 3g per 100g and per normal serving.

But many manufacturers do not follow the voluntary guidelines. The term 'light' is often used to indicate a reduced-fat product, but this is not included in the voluntary guidelines.

The law which currently governs nutritional claims made on packaged foods is inadequate. There are no detailed rules covering how and when claims about fat, saturates, sugars, fibre and sodium can be made. Also, food labelling regulations say a food's ingredients must be listed in order of weight on the packaging. But a product can be named after any one of its ingredients even if that ingredient is way down the list (see 'Food labelling', above).

Claims are covered only by general legislation which states that they have to be true and not misleading. Government guidelines exist but because they are merely guidelines they are not legally enforceable. If you think that the packaging or other advertising makes misleading health claims, report the claims to the ASA.

### Product recalls

If a food producer issues a warning about a product or recalls the item from retailers the clear implication is that it is defective. Product recalls are usually advertised in the press, and *Which?* magazine routinely publishes product recalls. Food recalls are fortunately quite rare. If you have such a product you should contact the manufacturers immediately. If it is already too late and damage has been done, ask the manufacturer for compensation (see below).

There are few guidelines for most domestic product recalls, but for foods and medicines special rules exist to ensure that unsafe products are taken off the market as soon as possible. No one can force a manufacturer to initiate a product recall although the government can make a manufacturer issue a notice warning people about an unsafe product. This is usually enough of a threat to make most manufacturers institute a recall.

## How to make a food-related complaint

If you have a serious problem with a food you have bought, you need to make your complaint quickly, effectively – and to the right person or organisation.

- Act quickly. If you discover a defect: go straight back to the shop, or, if that is inconvenient, write. If you are ill, get medical evidence immediately, otherwise you will not be able to show what caused your illness.
- Know your rights. This chapter explains what you are entitled to so you can let the person to whom you are complaining know the legal basis of your claim.

- Target your complaint. Write, or insist on speaking, to someone in authority – the manager of a local branch of a supermarket chain, for example, or the managing director of the company. Do not vent your anger on the telephonist or the cashier: he or she may not have the authority to take any action.

- Keep a record of your action. Even if you complain in person, or by phone, make sure you record what was said and when, together with a note of the name and position of the person with whom you dealt.

- Unless your complaint is resolved immediately, follow up by letter and keep a copy. By sending your letter to a named individual you reduce the risk of it being passed round the organisation and perhaps being ignored or lost. Use a heading and any reference given by the trader every time you write.

- When you write, stick to the point. Keep the letter brief, and set out the facts in short paragraphs. This will help your claim far more than an angry or emotional letter or one that includes personal remarks. But be firm. Quote the relevant legislation, such as the Sales of Goods Act 1979 if you are complaining about unsatisfactory quality in food, or the Supply of Goods and Services Act 1982 in respect of inadequate services: this will show that you are aware of your rights and you mean business. State what redress you want: money back, a replacement, or other financial compensation, for example. Give a reasonable deadline for a response, such as seven or 14 days. Use recorded delivery and keep copies of all documentation in addition to the letter itself.

- Get evidence – anything you can obtain that will support your claim: receipts, invoices, brochures, contract terms and conditions, advertisements, estimates, bills, statements from witnesses, photographs of damage, etc., and expert medical evidence if appropriate.

- Be persistent and don't be fobbed off. If you are not happy with the response to your complaint, or if you have had no response at all, write again. Do not fall prey to attempts to fob you off with less than you're entitled to.
  – *'We don't give refunds'* If you have bought food that is not of satisfactory quality, or not as described, say, you are entitled to a refund if you act quickly enough. Notices saying 'No refunds' are

against the law, so do not be deterred by them (report any you see to your local TSO)

– *'You caused the problem, not us'* Do not be put off by this. For example, as long as you follow the storage or cooking instructions the trader cannot blame you if food goes off prematurely or spoils in the cooking process

– *'It's not our problem. Try the manufacturer'* If you have bought food which turns out to be unsatisfactory, it is up to the retailer to deal with it, not the manufacturer. But all too often the retailer will try to pass you on to the manufacturer. Do not accept this: tell the retailer that it is his legal responsibility to address the problem

– *'No refunds on sale items'* If you buy food in a sale or at a knock-down price you still have your normal rights. So if your supermarket sells off dented cans of soup or cuts the price of food nearing or just past its 'best before' date you are still entitled to food of satisfactory quality, fit for consumption, and so on (see pages 149–50 and 159). But you cannot complain about any problems that were specifically pointed out to you or which you should have spotted before buying, such as the dent in the can, or the fact that the 'best before' date has passed.

- Be reasonable. Be prepared to come to a compromise if you receive a fair offer, even if it is not exactly what you wanted. But be warned that once you have accepted an offer of compensation you cannot ask for more later, so it is particularly important to take advice on your rights if you have been injured or made ill by food before deciding whether any offer is acceptable.
- Follow the right complaints procedure. If you cannot come to an agreement or if there is a breach of the criminal law it is time to let somebody else decide on the rights and wrongs of your complaint. Trading Standards and Environmental Health departments are the main bodies which deal with the criminal law. If the trader is a member of a trade association which operates a code of practice, you may be able to obtain help from it.

Trading Standards Departments, Environmental Health Departments, Citizens Advice Bureaux and Law Centres can offer help in a variety of ways, and the council officials can prosecute traders when necessary. But if you want to claim compensation for any loss or injury, or if your attempts fail to resolve the dispute with the trader on an amicable basis, you may have to consider taking legal action. In almost all cases you will have the option of going to court. The small claims court is part of the county court (Sheriff Court in Scotland) and offers a cheap and relatively informal way of dealing with fairly straightforward cases. If you have reached the end of your negotiations, the final stage before starting court action is to send a 'letter before action', telling the other party that unless you receive redress within a specified period (usually seven days) you will take the matter to court.

And remember, making a complaint is not just about the possibility of getting compensation, it is also about preventing something similar happening to other people, particularly in the case of illness or injury.

# WEIGHTS, MEASURES AND EGG SIZES

SINCE 1 October 1995, all pre-packed foods of variable weight (for example, fruit, vegetables, cheese, fish and meat) sold in the UK have had to have their weight marked in metric and be priced per kilogram (kg). The imperial equivalent can still be given as long as the metric information is more prominent.

Loose foods do not have to change until the year 2000. So at a deli counter or market stall you should still be able to ask for cheese by the ounce, labelled with price per pound. But pre-packed cheese will be marked in metric weight and labelled in price per kg. So it will not be all that easy to compare prices.

All the major supermarkets are distributing leaflets and displaying conversion charts to help their customers get used to the change. Generally, metrication should not lead to any price changes in foods. The doorstep pinta and the pint glass in the pub are to stay with us for the foreseeable future.

The information that follows is substantially reproduced from the Guild of Food Writers' leaflet entitled *Cooking in Metric*.

## Going metric

Measurements in British recipes have been given in imperial and metric units for over two decades. Many English-speaking countries such as Canada, South Africa, Australia and New Zealand now use metric-only measures. In line with EU legislation more and more foods in the UK are sold in metric units.

So now it makes sense to cook in metric. If you can count in 5s, 10s and 100s, you will find cooking in metric is just as easy.

Weight measurements are in grams (g) and kilos (kg). Volume and liquid measures are in ml (millilitres) and litres. Linear measures are in millimetres (mm), centimetres (cm) and metres (m). There are 1,000g in 1 kg (or kilo), 1,000ml in 1 litre, and 1,000mm in 1 metre.

Sometimes liquid measures also described in cl (centilitres), e.g. on a bottle of wine, or in dl (decilitres). There are 100cl to a litre and 10dl to a litre.

Measuring in metric is easy as long as your scales and measuring jugs carry both sets of measurements (most have done so for many years), while metric linear measures are already on rulers and tape measures.

To cook in metric, do not try to convert to or from imperial recipes. It could cause confusion, and most have been carefully worked out so that ingredients work with each other under one system or another (metric or imperial, but not a mixture of both). If a recipe calls for, say, 100g sugar and 300ml milk, then simply look at the relevant metric unit on your scales or measuring jug. Oven temperatures have been in Celsius (centigrade) for decades.

Cake tins and bowl sizes are generally expressed in cm and litres:. for example, a 20cm sandwich cake tin (approximately 8-inch) or 1-litre pudding basin (approximately 1¾-pint). If in doubt, simply measure your existing tins and bowls and note the metric size. There is no need to buy new ones.

If you have favourite old imperial recipes, there is no need to convert them into metric. You already have dual-measurement scales and jugs, so follow your recipes as before. You may find that as all foods such as butter, flour, sugar etc. are sold in metric units you might have some food left over after measuring, but that is all.

For jams and jellies, and in the absence of specific recipes, as a general rule you now think in terms of 500g sugar to each 500ml (½ litre) of fruit pulp or juice instead of the old guideline of 1 lb sugar to each pint.

Meat is now widely sold in decimal divisions of a kilo. So 500g of meat will be expressed as 0.5kg. To roast a 1.5kg joint, simply multiply the cooking time in the chart below × 1.5, or for a 2.2kg joint × 2.25.

*Oven temperature: gas 4, 180°C, 350°F*

| Beef: | rare | 35 minutes per 1kg + 20 minutes |
| | medium | 55 minutes per 1kg + 25 minutes |
| | well done | 65 minutes per 1kg + 30 minutes |

Pork:     65 minutes per 1kg + 30 minutes

Lamb:    55 minutes per 1kg + 35 minutes

Chicken:   50 minutes per 1kg + 20 minutes

| Whole salmon: | up to 2.5kg | 20 minutes per 1kg |
| | over 2.5kg | 16 minutes per 1kg |

## Metric and imperial equivalents

**Weight/solids**

| | | | |
|---|---|---|---|
| 15g | ½ oz | 500g | 1 lb 2 oz |
| 25g | 1oz | 600g | 1 lb 4 oz |
| 40g | 1½ oz | 700g | 1 lb 9 oz |
| 50g | 1¾ oz | 750g | 1 lb 10 oz |
| 55g | 2 oz | 900g | 2 lb |
| 75g | 2¾ oz | 1kg | 2 lb 4 oz |
| 85g | 3 oz | 1.2kg | 2 lb 12 oz |
| 100g | 3½ oz | 1.5kg | 3 lb 5 oz |
| 115g | 4 oz | 2kg | 4 lb 8 oz |
| 125g | 4½ oz | 2.25kg | 5 lb |
| 140g | 5 oz | 2.5kg | 5 lb 8 oz |
| 150g | 5½ oz | 3kg | 6 lb 8 oz |
| 175g | 6 oz | | |
| 200g | 7 oz | | |
| 225g | 8 oz | **Volume/liquids** | |
| 250g | 9 oz | 15ml | ½ fluid ounce |
| 275g | 9¾ oz | 30ml | 1 fl oz |
| 280g | 10 oz | 50ml | 2 fl oz |
| 300g | 10½ oz | 100ml | 3½ fl oz |
| 325g | 11½ oz | 125ml | 4 fl oz |
| 350g | 12 oz | 150ml | 5 fl oz (¼ pint) |
| 400g | 14 oz | 200ml | 7 fl oz |
| 425g | 15 oz | 250ml (¼ litre) | 9 fl oz |
| 450g | 1 lb | 300ml | 10 fl oz (½ pint) |

| | | | |
|---|---|---|---|
| 350ml | 12 fl oz | 750ml (¾ litre) | 1½ pints |
| 400ml | 14 fl oz | 1 litre | 1¾ pints |
| 425ml | 15 fl oz (¾ pint) | 1.2 litres | 2 pints |
| 450ml | 16 fl oz | 1.5 litres | 2¾ pints |
| 500ml (½ litre) | 18 fl oz | 2 litres | 3½ pints |
| 600ml | 1 pint (20 fl oz) | 2.5 litres | 4½ pints |
| 700ml | 1¼ pints | 3 litres | 5¼ pints |

Metric spoon sizes are 15ml, 10ml and 5ml, all of which are close to existing tablespoon, dessertspoon and teaspoon sizes, so you can continue to use your usual spoons for measuring ingredients.

## Egg sizing

From late 1995, eggs have been sold in new, easier-to-understand sizes – small, medium, large and extra large – rather than in the sizes 0-7 used for several years previously. Many supermarkets also show a minimum net weight on the packaging. The old and new egg sizes are:

| OLD SIZE | NEW SIZE |
|---|---|
| 0 | Extra large |
| 1 and 2 | Large |
| 3 and 4 | Medium |
| 5 & above | Small |

Many recipes list size 3 eggs in their ingredients lists: use 'medium' eggs under the new system.

# Meat cuts and cooking methods

THIS SHORT section shows where the various cuts of meat come from, for beef, lamb and pork, and which cooking methods can be used for each of them.

## Beef

**Note** As we go to press, it has been established that there is an unquantifiable risk in eating beef and beef products. Consumers' Association has advised those who wish to avoid the risk from BSE-infected beef altogether to cut out beef and beef products from their diet. The highest risk is from beef products, such as burgers, pies and sausages. There is less risk from beef bought in joints or as steaks, for example (that is, muscle tissue), and from organic beef, but no beef can currently be guaranteed BSE-free.

**Braising steak (chuck, blade, shoulder, thick rib)** Braise or casserole, stew

**Brisket (on the bone or boned and rolled)** Pot-roast, braise, boil

**Chuck or stewing steak** Stew, casserole

**Clod or sticking (cut found between neck and shin) or neck** Stew, casserole

**Flank** Pot-roast, braise, boil

**Flash-fry cuts (top rump, thick rib, silverside)** Grill, fry, stir-fry

**Fillet (whole or fillet steak)** Whole: roast or microwave. Steak: grill, fry, barbecue

**Leg** Stew, casserole

**Minced/ground beef** Fry, microwave, grill, barbecue as burgers, bake

**Rib** Fore-rib (rib roast, whole or boned and rolled): roast, microwave. Top and back: slow-roast. Wing or prime: roast. Thin rib and thick ribs (boned and rolled): braise, pot-roast

**Rump, top (also known as thick flank)** Pot roast, slow roast

**Shin** Stew, casserole

**Skirt** Stew, casserole

**Steaks** Chateaubriand: grill, fry. Entrecôte: grill, fry. Mignon (also known as filet mignon or tournedos): grill, sauté. Porterhouse: grill, fry. Rump: grill, fry, stir-fry, barbecue. Sirloin: grill, fry, stir-fry, barbecue. T-bone: grill, fry

**Silverside** Roast, braise, pot roast, boil (salted)

**Sirloin (on bone or boned and rolled)** Roast, microwave

**Stewing steak (shin, leg, neck or clod)** Stew, casserole, braise

**Stir-fry beef (topside, sirloin or rump)** Stir-fry

**Topside (also sold sliced)** Roast, microwave, pot-roast, grill, stir-fry, fry when sliced

**Thick flank (also sold sliced)** Roast, pot-roast, fry when sliced

## Lamb

**Best end of neck (rack of lamb)** Braise, roast

**Breast (can be boned, stuffed or rolled)** Roast, braise, casserole

**Chops (loin, leg or chump)** Grill, fry, braise, barbecue, microwave, roast

**Chump joint (may be boned and rolled)** Roast

**Cutlets** Best end of neck: grill, fry, microwave. Middle neck: stew, braise

**Cubed lamb (meat from shoulder, leg or chump)** Casserole, grill or barbecue as kebabs, stir-fry

**Leg (gigot) joints** Whole: roast, braise, boil, microwave. Fillet (upper part): roast, microwave. Knuckle (lower part): roast, microwave

**Loin (on bone or boned)** Roast

**Middle neck** Stew, braise, casserole, fry if sliced

**Minced/ground** Microwave, fry, grill or barbecue as burgers, bake

**Neck fillet** Grill, stir-fry, barbecue, braise, fry, roast

**Noisettes (slices of loin or best end)** Grill, fry

**Rib (cut into best end of neck, middle neck and scrag)** Roast, braise

**Riblets (short rib cuts from breast)** Grill, roast

**Saddle** Roast

**Scrag end of neck (sold cubed)** Stew, soup, broth

**Shoulder (also sold boned and rolled; half-shoulder sold as blade or knuckle end)** Roast, braise

# Pork

**Belly pork (also known as streaky, draft or flank pork)** Roast, braise, diced for stews, sliced for grilling, barbecue

**Blade (on bone or boned, stuffed and rolled)** Roast

**Chump** Roast

**Chops/steaks** Chump and loin: grill, fry, bake, roast, microwave, barbecue. Spare rib: grill, fry, braise, casserole. Leg and shoulder: grill, fry, barbecue, stir-fry, casserole

**Cubed pork** Grill or barbecue as kebabs, stir-fry, casserole

**Escalopes (very lean meat, sliced)** Grill, fry, stir-fry, sauté

**Fillet or tenderloin** Roast, braise, grill, fry, sauté, microwave

**Fillet half leg (top end of hind leg)** Roast

**Hand (lower part of shoulder)** Roast

**Hand and spring (also known as shoulder or runner)** Roast, braise, microwave, pot-roast

**Knuckle** Roast, boil, stew

**Leg (gigot) joints** Fillet end: roast, microwave. Shank (knuckle) end: pot-roast

**Leg and shoulder steaks** Grill, fry, microwave, foil-bake

**Loin** Roast, pot-roast, microwave

**Minced/ground** Bake, fry, microwave

**Neck end (large joint often divided into blade and spare rib)** Roast, pot-roast

**Spare-rib joint (can be boned, stuffed and rolled)** Roast, braise, stew

**Spare ribs and riblets (also known as barbecue ribs, Chinese-style ribs, American-cut ribs)** Grill, barbecue, bake, roast

**Stir-fry pork** Stir-fry

**Beef cuts**

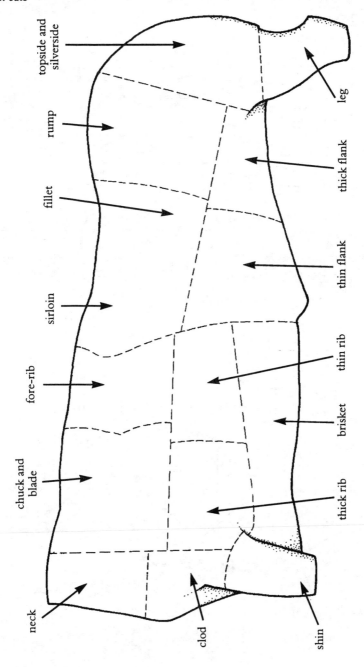

**Lamb cuts**

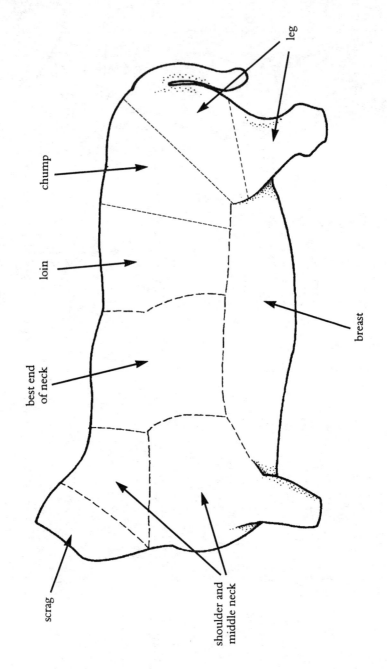

**Pork cuts**

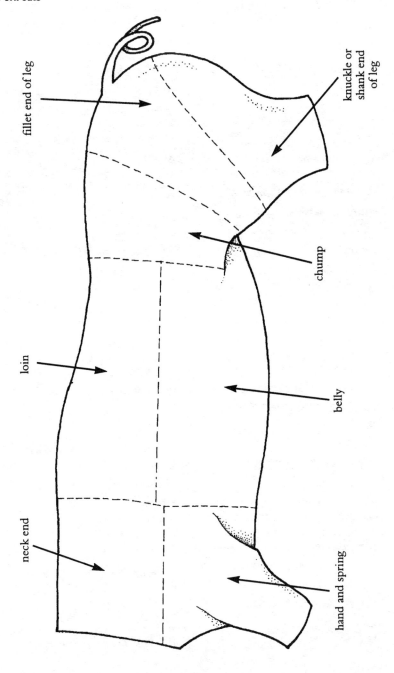

fillet end of leg

knuckle or shank end of leg

chump

loin

belly

neck end

hand and spring

# BRITISH FRUIT CALENDAR

## January

**Apples** Dessert: Cox's Orange Pippin, Laxtons Superb, Spartan Cooking: Bramleys Seedling
**Pears** Dessert (Comice, Conference)
**Rhubarb** (forced)

## February

**Apples** Dessert: Cox's Orange Pippin, Crispin, Laxtons Superb. Cooking: Bramleys Seedling
**Pears** Dessert: Comice, Conference
**Rhubarb** (forced)

## March

**Apples** Dessert: Cox's Orange Pippin, Crispin, Laxtons Superb Cooking: Bramleys Seedling
**Pears** Dessert: Comice, Conference
**Rhubarb** (main crop)

## April

**Apples** Dessert: Cox's Orange Pippin, Crispin, Laxtons Superb Cooking: Bramleys Seedling
**Gooseberries**
**Rhubarb** (main crop)

## May

**Apples** Dessert: Cox's Orange Pippin
Cooking: Bramleys Seedling
**Gooseberries**
**Rhubarb** (main crop)
**Strawberries**

## June

**Apples** Cooking: Bramleys Seedling
**Gooseberries**
**Rhubarb** (main crop)
**Strawberries**

## July

**Apples** Cooking: Bramleys Seedling
**Bilberries**
**Blackcurrants**
**Blueberries**

Cherries
Gooseberries
Loganberries
Plums
Raspberries
Redcurrants
Strawberries
Whitecurrants

## August

**Apples** Dessert: Discovery
Cooking: Early Victory,
Grenadier
Bilberries
Blackberries
Blackcurrants
Blueberries
Gooseberries
Greengages
Loganberries
**Pears** Dessert: Williams
Plums
Raspberries
Redcurrants
Strawberries
Whitecurrants

## September

**Apples** Dessert: Cox's Orange
Pippin, Discovery, Worcester
Pearmain
Blackberries
Blueberries
Crab apples

Damsons
Greengages
**Pears** Dessert: Conference
Plums

## October

**Apples** Dessert: Cox's Orange
Pippin, Egremont Russet,
Spartan, Worcester Pearmain
Cooking: Bramleys Seedling
Blackberries
Crab apples
**Pears** Dessert: Comice,
Conference, Packhams
Quince

## November

**Apples** Dessert: Cox's Orange
Pippin, Crispin, Egremont
Russet, Laxtons Superb, Spartan,
Worcester Pearmain
Cooking: Bramleys Seedling
**Pears** Dessert: Comice,
Conference, Packhams
Quince

## December

**Apples** Dessert: Cox's Orange
Pippin, Crispin, Spartan.
Cooking: Bramleys Seedling
**Pears** Dessert: Comice,
Conference
Rhubarb (forced)

# BRITISH VEGETABLE CALENDAR

## January

Artichokes (Jerusalem)
Beetroot
Brussels sprouts
Cabbages (white, winter, savoy, red)
Carrots
Cauliflower
Celeriac
Kohl rabi
Leeks
Mushrooms
Onions
Parsnips
Potatoes
Shallots
Spring greens
Swede
Turnips

## February

Artichokes (Jerusalem)
Brussels sprouts
Cabbages (white, winter, savoy, red)
Carrots
Cauliflower
Leeks
Mushrooms
Onion
Parsnips
Potatoes
Shallots
Swede
Turnips

## March

Artichokes (Jerusalem)
Broccoli (sprouting)
Brussels sprouts
Cabbages (white, winter, savoy, red)
Carrots
Cauliflower
Leeks
Mushrooms
Onions
Parsnips
Potatoes
Spinach
Swede
Tomatoes
Turnip

## April

Broccoli (sprouting)

**Cabbages** (winter, red, white, greens)
**Carrots**
**Cauliflower**
**Chicory**
**Cucumber**
**Leeks**
**Mushrooms**
**Potatoes**
**Spinach**
**Swede**
**Tomatoes**
**Watercress**

## May

**Asparagus**
**Broccoli** (sprouting, Cape)
**Carrots**
**Cauliflower**
**Chinese leaves**
**Cucumber**
**Leeks**
**Mushrooms**
**Potatoes**
**Radishes**
**Spinach**
**Spring greens**
**Tomatoes**
**Watercress**

## June

**Asparagus**
**Beetroot**
**Beans** (broad)
**Broccoli**
**Carrots**

**Cauliflower**
**Chicory**
**Chinese leaves**
**Courgettes**
**Cucumber**
**Endive**
**Lettuce**
**Mushrooms**
**Peas**
**Potatoes** (new)
**Radishes**
**Spinach**
**Spring onions**
**Sprouting broccoli**
**Tomatoes**
**Turnips**
**Watercress**

## July

**Beetroot**
**Beans** (broad, runner)
**Broccoli**
**Carrots**
**Cauliflower**
**Celery**
**Chicory**
**Chinese leaves**
**Courgettes**
**Cucumber**
**Endive**
**Fennel**
**Lettuce**
**Mushrooms**
**Peas**
**Peppers**
**Potatoes** (new)
**Radishes**
**Tomatoes**

Turnips
Watercress

## August

Artichokes (globe)
Beetroot
Beans (broad, French, runner)
Broccoli
Carrots
Cauliflower
Celery
Chicory
Chinese leaves
Courgettes
Cucumber
Endive
Fennel
Kohl rabi
Leeks
Lettuce
Marrow
Mushrooms
Peas
Peppers
Potatoes (new)
Radishes
Tomatoes
Turnips

## September

Beans (broad, French, runner)
Broccoli
Carrots
Cauliflower
Celery
Chinese leaves

Courgettes
Cucumber
Endive
Fennel
Kohl rabi
Leeks
Lettuce
Marrow
Mushrooms
Onions
Parsnips
Peppers
Potatoes
Pumpkins
Radishes
Spinach
Sweetcorn
Tomatoes
Turnips

## October

Broccoli
Brussels sprouts
Cabbages (white, savoy, red)
Carrots
Cauliflower
Chicory
Chinese leaves
Endive
Fennel
Kohl rabi
Leeks
Lettuce
Marrow
Mushrooms
Onions
Parsnips

Potatoes
Pumpkins
Shallots
Spinach
Sweetcorn
Tomatoes
Turnips

## November

Artichokes (Jerusalem)
Brussels sprouts
Cabbages (white, savoy, red)
Carrots
Cauliflower
Chinese leaves
Kohl rabi
Leeks
Mushrooms
Onions
Parsnips

Potatoes
Shallots
Turnips

## December

Artichokes (Jerusalem)
Brussels sprouts
Cabbages (white, savoy, red)
Carrots
Cauliflower
Celeriac
Kohl rabi
Leeks
Mushrooms
Onions
Parsnips
Potatoes
Shallots
Swede
Turnips

# HERBS AND SPICES

HERBS AND spices add flavours of their own to foods as well as enhancing the natural flavours of food, but the aim should be for them to complement rather than dominate the final flavour.

Fresh herbs are preferable to dried as they generally have a better flavour and colour. Frozen herbs are a good alternative. Some dried herbs have a much better flavour than others. Many make excellent store-cupboard items.

Many fresh herbs can now be bought all year round and many can be grown in tubs or pots in the garden. Supermarkets also sell some varieties as pot herbs with roots intact. These, correctly stored, should last for several days.

Fresh herbs should always be washed gently or wiped before use to remove any soil or dust particles. Fresh herbs may be stored in a jug of water at room temperature or in an unsealed polythene bag in the refrigerator.

Dried herbs should be stored in airtight containers in a cool, dark, dry cupboard. Fresh herbs may also be dried at home: simply tie the herbs in bunches and hang them upside-down in a dry, warm, well-ventilated place until dry.

Spices are also available in fresh and dried forms, whole or ground. It is always best to grind your own spices as and when you need them as the flavour will be much fresher, stronger and more aromatic.

Ground spices may lose their flavour and colour quickly as their essential oils fade. It is worth investing in a pestle and mortar to grind your own spices. Alternatively, use a coffee grinder (kept only for grinding spices) or a peppermill.

Store spices in airtight, dark glass containers in a cool, dark, dry cupboard to keep them at their best.

# Herbs

| Food | How to choose | Storage | Freezing |
|------|---------------|---------|----------|
| **Basil**<br>Many types, all of them suitable for culinary purposes. Warm flavour with spicy overtones of aniseed. Available fresh and dried. | Check 'best before' date. Dried: should retain aniseed flavour. Fresh: should have good, fresh green colour. | Dried: keep in airtight container in a cool, dark, dry place. Fresh: best used as fresh as possible. Keep for short period in plastic bags in refrigerator at 0-5°C or in jug of water. Use by recommended date. | Chop leaves finely and place in ice-cube trays with a little water. After freezing, store cubes in sealed plastic bags in the freezer at −18°C. |
| **Bay**<br>Spicy flavour, a little like fresh ground pepper. Available fresh and dried. | Check 'best before' date. Fresh leaves should be dark green and glossy with no blemishes. | Dried: keep in airtight container in a cool, dark, dry place. Fresh: best used at once; otherwise, store in a plastic bag in refrigerator at 0-5°C for a few days. Use by recommended date. | Not applicable. |
| **Chervil**<br>Delicate parsley-like flavour. Available fresh and dried. | Check 'best before' date. Fresh: should have a good colour. | Dried: keep in airtight container. Fresh: best used at once; otherwise, store in a plastic bag in refrigerator at 0-5°C. Use by recommended date. | As basil. |
| **Chives**<br>Delicate onion flavour. Available fresh and dried. | Check 'best before' date. If fresh, should be bright green. | Dried: keep in airtight container in a cool, dark, dry place. Fresh: store in an airtight container in refrigerator at 0-5°C. Use by recommended date. | As basil. |
| **Coriander**<br>All parts may be used for cooking. Flavour varies from the faint anise of the leaves to the citrus peel flavour of the seeds. Available fresh and dried. Seeds are available dried in whole or ground forms. | Check 'best before' date. Fresh: leaves should have good green colour. Whole seeds: should be light brown and even in size. | Dried: keep in airtight container in a cool, dark, dry place. Fresh: best used as fresh as possible. Store in a plastic bag wrapped in wet paper towels. Use by recommended date. | As basil. |

| Preparing and cooking | Typical uses | Alternative forms |
|---|---|---|
| Fresh leaves are better torn than cut. Add at last minute to cooked dishes. | Has special affinity with tomatoes. Use in salads, pesto, as a garnish, for flavouring savoury dishes (pasta sauces, pizzas, other Mediterranean dishes), basil oil and vinegar. | Also available as pot herb, freshly minced in a jar, frozen. |
| Usually used dry. Fresh leaves give more flavour if torn or broken. Flavour strengthens as cooking time increases. Remove whole leaves before serving. | Soups, stews and sauces. Sweet dishes such as milk puddings or custards. As a garnish, kebabs. | Ground. |
| Add at last minute to avoid flavour loss. | Chicken and egg dishes, fish, garnish, sauces, soups, vegetables. | |
| Cut with scissors or a sharp knife to avoid bruising. Add at last minute to cooked dishes. May also be blanched and used for tying small bundles of vegetables together. | Chicken dishes, egg dishes, such as omelettes, fish dishes, garnish, potato dishes such as potato salad, salads, soups. | Frozen. As pot herb. Chive flowers used in salads or as decoration. |
| Add fresh leaves at end of cooking. | Leaves: add to chicken and fish dishes, curries, rice, and tomato dishes, as a garnish.<br>Seeds: in chutney and pickles, curries, pork dishes. Also in fruit crumbles (e.g. rhubarb), cakes, pasties, vegetables. | Pot herb. Freshly minced in a jar. |

| Food | How to choose | Storage | Freezing |
|------|---------------|---------|----------|
| **Dill** Spicy anise-like flavour and aroma; rather sweet. Available as fresh and dried leaves and as whole or ground seeds (dried). | Check 'best before' date. Fresh: should have a good green colour. | Dried: keep in airtight container in a cool, dark, dry place. Fresh: store in a plastic bag in refrigerator at 0-5°C. Use by recommended date. | As basil. |
| **Lemon balm** Member of mint family. Crushed leaves give off a strong lemon fragrance. Available fresh and dried. | Check 'best before' date. Fresh: should have a good light green colour. | Dried: keep in airtight container in a cool, dark, dry place. Fresh: best used at once; otherwise, store in a plastic bag in refrigerator at 0-5°C for a few days. Use by recommended date. | Not applicable. |
| **Lemon grass** Lemon-flavoured herb. Available fresh and dried. | Check 'best before' date. Dried ground lemon grass should have a fresh lemony flavour and be pale green in colour. | Dried: keep in airtight container in a cool, dark, dry place. Fresh: store in a plastic bag in refrigerator at 0-5°C for a few days. Use by recommended date. | Not applicable. |
| **Marjoram/oregano** Closely related herbs, of which marjoram has the more delicate flavour. Available fresh and dried. | Check 'best before' date. Fresh: should have leaves of good colour and even size. | As above. | As basil. |
| **Mint** Many types e.g. apple mint, lemon mint, peppermint, spearmint. Available fresh and dried. | Check 'best before' date. Fresh: should have a good green colour. Dried: should have a strong mint flavour and bright green colour. | As lemon balm. | As basil. |
| **Parsley** Two types: regular and flat leaf (the latter has more flavour). Available fresh and dried. | Check 'best before' date. Fresh: should have a good green colour. Dried: should have a mild aroma and good colour. | Dried: keep in airtight container in a cool, dark, dry place. Fresh: best kept with cut ends in cold water. Store in a plastic bag in refrigerator at 0-5°C or sprinkled with water and wrapped in paper towels. Use by recommended date. | As basil. |

| Preparing and cooking | Typical uses | Alternative forms |
|---|---|---|
| Add fresh dill to dishes just before serving. Use dried dill generously as it has little flavour. | Seeds: bread, cakes, fish dishes, pickles, soups, stews.<br>Leaves: cream cheese, eggs and fish dishes, garnish, marinades, potato salad, salads, soups. | Pot herb. Dill sauce. |
| Chop leaves finely before use. Use whole or torn leaves for infusing. | Fruit drinks, fruit salad, garnish, herbal tea, infused in milk for custards, omelettes, salads, sauces, soups. | Pot herb. |
| Use fresh stalks either whole or chopped. Infuse leaves in hot water for herbal tea. | Chicken casseroles, curries, meat stews, salads, sea food, soups, stir-fries, Thai dishes. | |
| Add fresh herb at the end of cooking. | Greek and Italian dishes, pizzas, poultry, sea food and salad dressings, tomato-based sauces. | Pot herb. |
| Use chopped in dishes or whole as a garnish. | With boiled new potatoes, in fruit salads, as garnish, herb tea, lemon-based puddings such as lemon mousse, mint sauce with lamb, salads. | Pot herb, mint jelly, mint sauce. |
| Add to food at end of cooking. | Deep-fried, as garnish, in meat and poultry dishes, parsley sauce, pasta dishes, salads, parsley and thyme stuffing. | Pot herb. |

| Food | How to choose | Storage | Freezing |
|------|---------------|---------|----------|
| **Rosemary** Pinewood perfume and a strong bitter-sweet flavour. Highly aromatic. Available fresh and dried. | Check 'best before' date. Fresh: should have dark green needle-like leaves with a silver downside. | Dried: keep in airtight container in a cool, dark, dry place. Fresh: store in a plastic bag in refrigerator at 0-5°C or with stems in cold water. Use by recommended date. | As basil. |
| **Sage** Many varieties, including common purple and golden. Subtle, sweet pinewood flavour. Available fresh and dried. | Check 'best before' date. Fresh: silver/grey in colour. | Dried: keep in airtight container in a cool, dark, dry place. Fresh: store in a plastic bag in refrigerator at 0-5°C for a few days. Use by recommended date. | As basil. |
| **Tarragon** Strong sweet and spicy flavour. Available fresh and dried. | Check 'best before' date. Fresh: should have greyish-green leaves with a peppery scent. Dried: should have an anise-like flavour. | As lemon grass. | As basil. |
| **Thyme** Many varieties. Aromatic, with strong, pungent flavour. Available fresh and dried. | Check 'best before' date. Fresh: should have leaves of a good green colour. Dried: should have a strong fresh aroma and contain little or no twig or stalk. | As lemon grass. | As basil. |

## Spices and flavourings

| | | | |
|------|---------------|---------|----------|
| **Allspice** Dried unripe berry of a tree, dark reddish-brown in colour. Berries are the size of large peppercorns with a fragrant aroma. Tastes of a mixture of cinnamon, nutmeg and cloves. | Check 'best before' date. Choose unbroken, firm berries, not too dry. Aroma is good indication of quality. | In an airtight container in a cool, dark, dry place. Whole: use within a year. Ground: use within 6 months. Use by recommended date. | Freeze in dishes. |
| **Caraway** Dried seeds of a herb, dark brown in colour with warm, slightly bitter aniseed flavour and pungent aroma. Available dried as seeds and ground. | Check 'best before' date. Seeds should be unbroken, firm, not too dry. Aroma is good indication of quality. | In an airtight container in a cool, dark, dry place. Use within a year. Use by recommended date. | Freeze in dishes. |

| Preparing and cooking | Typical uses | Alternative forms |
|---|---|---|
| Finely chop or crush fresh leaves before use. Use in moderation. | Herb butter, meat dishes (especially lamb), pizza, tomato-based sauces. Also sweet dishes such as fruit salad or jelly. | Flowers. Pot herb. |
| Use sparingly. Use chopped or as whole leaves. | Cheese dishes, kebabs, meats such as pork, sage and onion stuffing, sausage dishes, sautés, stews, tomato-based sauces. | Pot herb. |
| Use as whole leaves, chopped or shredded. | Chicken dishes, egg dishes, potato dishes, salads, sauces, soups, vegetables. | Pot herb. Preserved in vinegar. |
| Retains flavour well in long cooking. | Casseroles, cheese dishes, chicken dishes, fish dishes, parsley and thyme stuffing, sauces, soups, stews. | Pot herb. |
| A good all-round spice. Enhances flavour of many other spices. Use in sweet or savoury dishes. Often used with cinnamon and cloves. | Chutneys, ketchup, marinades, meat, fish and poultry dishes, mulled drink, pickles. | Pickling spice ingredient. |
| Use in both sweet and savoury dishes. Seeds need long cooking to release flavour. | Apple pie, breads, cakes, meat dishes, e.g. with pork, duck or goose, pickles, sauerkraut, sausages, vegetable dishes. | In bread and crackers. |

| Food | How to choose | Storage | Freezing |
|------|---------------|---------|----------|
| **Cardomom** Dried unripe fruit of a bush. Seeds are encased in a pod and have a citrus-like flavour. Available dried as pods, loose seeds or ground. | Check 'best before' date. Best bought as whole pods because loose seeds and ground cardamom quickly lose their flavour. Seeds should be unbroken, firm, not too dry. Aroma is good indication of quality. | As allspice. | Freeze in dishes. |
| **Celery seeds** Seeds of the celery plant, brown in colour with a pungent, almost bitter, celery flavour. Available whole or ground. | Check 'best before' date. Seeds should be unbroken, firm, not too dry. Aroma is good indication of quality. | In an airtight container in a cool, dark, dry place. Use within a year. Use by recommended date. | Freeze in dishes. |
| **Chillies (dried)** Dried pods of a plant of which there are many different varieties. Most common are deep, rich red colour. Heat and flavour vary considerably. Available whole, crushed, ground or flaked. | Check 'best before' date. Choose chillies with a good colour, unbroken and whole. | As allspice. | Freeze in dishes. |
| **Cinnamon** Dried rolled inner bark from a small tree. Soft brown in colour with a warm, rich, spicy aroma. Available as rolled sticks, quills or ground. | Check 'best before' date. Aroma is good indication of quality. | As allspice. | Freeze in dishes. |
| **Cloves** Dried unopened flower buds from a small tree. Warm reddish-brown colour and smell strong and sweet with a pungent flavour. Available whole and ground. | Check 'best before' date. Choose cloves that are bright reddish-brown in colour. Should snap cleanly and be unbroken, firm, not too dry. Aroma is good indication of quality. | As allspice. | Freeze in dishes. |

| Preparing and cooking | Typical uses | Alternative forms |
|---|---|---|
| Use to enhance both sweet and savoury dishes. The outer pod is not eaten. | Cakes, cooked fruit dishes, curries, ice-creams, pastries, pickles, puddings. Standard ingredient of garam masala. | |
| Crush seeds before use or sprinkle over vegetables before serving. | Chutneys, egg dishes such as omelettes, fish dishes, pickles, sauces, soups, sprinkled over bread, stews, stuffings, with tomatoes in pasta sauces. | Celery salt made from crushed seeds mixed with salt, fresh celery stalks and leaves. |
| Strength of flavour increases with cooking. | Chilli con carne and other Mexican dishes, Indian, Chinese and South East Asian dishes, pickles, piquant sauces, pizza and pasta sauces, sausages. Curry powder ingredient. | Canned, cayenne pepper, chilli oil, chilli sauce, chilli vinegar. Fresh chillis (many types) are available whole, minced, frozen, paprika, paste, pickled, preserved in oil, roasted, smoked. |
| Use in both sweet and savoury dishes. Use ground in baking. Remove whole sticks before serving. | Ground: apple pies and puddings, baked custards, biscuits, cakes, curries, meat stews especially lamb, puddings and stewed fruit. Whole sticks: chutneys, infused in liquid for spiced drinks such as mulled wine, pickles and sweet sauces. | |
| As for cinnamon. | Bread sauce, Chinese and Indonesian dishes, chutneys, desserts (especially apple-based), fruit cakes and puddings, mulled wine, stews, stocks, for studding meats (such as ham) or fruit. | Clove oil. |

| Food | How to choose | Storage | Freezing |
|------|---------------|---------|----------|
| **Cumin**<br>Dried seeds from an umbrella-shaped plant. Slightly bitter, warm, earthy, aromatic flavour. Available as whole seeds or ground. | Check 'best before' date. Aroma is good indication of quality. | As allspice. | Freeze in dishes. |
| **Fenugreek**<br>Herb producing seeds with a nutty, slightly bitter flavour and a strong, pungent aroma. Seeds available crushed, ground, dried, whole, fresh, sprouted. Leaves available fresh and dried. | Check 'best before' date. Seeds should be unbroken, firm, not too dry. Aroma is good indication of quality. | As allspice. | Freeze in dishes. |
| **Garlic**<br>Member of the onion family: a head or bulb of garlic is made up of several small cloves held together by an outer papery covering. Strong characteristic flavour and smell. Various types of different size, colour and flavour. Most common are white and pink-skinned. | Check 'best before' date. Choose plump, succulent bulbs with large cloves, bright and white in colour. Head of garlic should be firm, compact and not sprouting. Peeled cloves should be creamy-white in colour. Avoid soft, dark-coloured, mouldy, shrivelled or sprouting garlic. | Store in cool, dark, dry, airy place. Store bulbs in earthenware pot with air holes: will keep for several months like this. Whole strings of garlic may be hung in cool, dry place. If garlic begins to shoot, remove and discard green shoot completely before using, as it is very strong and bitter. Use by recommended date. | Freeze in dishes. Crushed garlic does not freeze well. Garlic butter freezes very well. |
| **Ginger**<br>The knobbly root of the ginger plant is beige in colour and has a refreshing, slightly hot and spicy flavour and warm sweet smell. It adds a rich, warming flavour to cooked food. Available fresh, dried, in slices or ground. | Check 'best before' date. Fresh: choose firm-fleshed roots which look fresh. Avoid soft, wrinkled ones. | Dried: store in a cool, dark, dry place in an airtight container. Use within a year.<br>Fresh: wrap in absorbent kitchen paper and keep in a plastic bag in the refrigerator at 0-5°C for up to 2 weeks. Use by recommended date. | Freeze in dishes. |

| Preparing and cooking | Typical uses | Alternative forms |
|---|---|---|
| | Couscous, curries, meat stews, Mexican dishes, pickles, tomato-based sauces. Often used with coriander. Standard ingredient of curry powder. | |
| Lightly roast seeds before use. | Chutneys, curries, pickles, stews and as a coating for fried food. | |
| Peel and finely chop cloves, slice or crush them using a garlic press, or use cloves whole; or use whole bulb – roasted with skin on, then peeled to serve. Quantity depends on personal taste. Use sparingly in combination with delicate flavours. For just a hint of garlic rub a clove round bowl before serving a salad. Never let garlic burn when frying it: it will become bitter. Cook it gently and quickly. | Add to savoury dishes as a flavouring, e.g. aïoli, garlic butter and bread, marinades, pasta sauces, pesto, salad dressings, soups. Cook and serve as a vegetable. | Dried, flakes, frozen, minced, pickled, powder, purée, salt, wood-smoked whole garlic bulbs. |
| Peel, using a potato peeler, and grate, slice or chop finely. Use in sweet or savoury dishes. Bruise dried ginger slices before using to release flavour. | Baked goods (cakes, sweet breads etc.), biscuits (e.g. gingerbread men), chutneys, curries, fruit dishes, meat and poultry stews, mulled wine, pickles, stewed or baked puddings. | Beer and cordials, candied, crystallised, freshly minced ginger, oil (used for flavouring wine), glacé, pickled, preserved in syrup. |

| Food | How to choose | Storage | Freezing |
|------|---------------|---------|----------|
| **Horseradish** Root is most commonly used part, although young leaves may be used in salads. Powerful, pungent, hot taste. | Look for firm, fat, unblemished roots. Avoid sprouting roots or ones which are slightly green in colour as they may be bitter. | Keeps for several weeks in salad drawer of refrigerator. | Freeze fresh root peeled and grated or shredded in sealed freezer bags. |
| **Mace and nutmeg** These, from the same tree, add warmth and sweetness to cooked dishes. Mace, available as blades or ground, is the bright red lacy covering surrounding the walnut-sized seed. The nutmeg, available whole or ground, is the kernel of the seed. | Check 'best before' date. Buy nutmeg whole and grate as needed. | Store in a cool, dark, dry place in an airtight container. Whole: stored correctly, will keep indefinitely. Ground: use within 6 months. Use by recommended date. | Freeze in dishes. |
| **Mustard** Dried seeds from the mustard plant. Strong, biting flavour and a fresh, clean aroma. Available as whole seeds, ground or crushed. | Check 'best before' date. Whole seeds should be unbroken, firm, not too dry. | As allspice. | Freeze in dishes. |
| **Nutmeg** See mace. | | | |
| **Olives** Small, firm fruit of the olive tree. Even when ripe, olives are bitter and inedible and need to be cured – in oil, water, brine or salt – when green and unripe or when black and ripe. Available green or black. | Check 'best before' date. | Olives packed in oil, brine or vinegar may be stored at a cool room temperature for many months if unopened. Canned and vacuum-packed olives keep unopened for many months; after opening, transfer to a bowl, cover and store in refrigerator at 0-5°C for up to 3 days. Refrigerate olives bought loose at 0-5°C for up to 3 days. Use by recommended date. | Freeze in dishes. |

| Preparing and cooking | Typical uses | Alternative forms |
| --- | --- | --- |
| Fresh: peel, then grate or shred the flesh, preferably in a food processor as grating by hand may irritate the skin and cause burning and watery eyes. Fresh horseradish loses its pungency quickly so prepare only small quantities. Dried horseradish flakes or powder may be reconstituted and used as fresh. | As a condiment (sauce or relish), particularly good with foods such as roast beef, smoked fish, eggs, chicken. | Horseradish sauce or relish, creamed horseradish. |
| Add freshly grated nutmeg at the end of cooking. | Biscuits and cakes, custards, milk puddings, sauces, seafood dishes, spinach dishes, stewed fruit, vegetables. | Mace oil. |
| Mix powder with water. Use sparingly. | Fish and cheese dishes, mayonnaise and salad dressings, relishes, pickles, soups, stews, vegetables. | Prepared or blended, plain or flavoured, smooth or wholegrain, in pickling spice, oil. |
| Use whole, halved, chopped or puréed. | As cocktail snack, in canapés, accompaniment to meat and fish, in braises, Mediterranean dishes, salade niçoise, olive bread, pizzas, salads, stews, tapenade. | Stuffed with pimiento, almonds, capers or onion. In marinade of oil flavoured with garlic and herbs. Olive oil, olive spread/pâté, tapenade. |

| Food | How to choose | Storage | Freezing |
|------|---------------|---------|----------|
| **Pepper**<br>Dried berries from a tropical trailing vine. Fresh, pungent flavour depending on type. The same plant produces black, white and green peppercorns. Available whole, crushed or ground. | Check 'best before' date. Buy whole peppercorns (unbroken, firm, not too dry) for best flavour and grind as needed. Aroma is good indication of quality. | In an airtight container in a cool, dark, dry place. Dried and whole: keep indefinitely. Ground: use within 6 months. Use by recommended date. | Freeze in dishes. |
| **Poppy seeds**<br>Edible seeds of the opium poppy. Crunchy, with nutty flavour and blue-grey (alternatively, pale yellow or brown) colour. Available whole and ground. | Check 'best before' date. | In an airtight container in a cool, dark, dry place. Use within a year or by recommended date. | Freeze in dishes. |
| **Saffron**<br>The dried stigmas from crocus flowers, with an aromatic honey-like flavour and deep, rich orange-red colour. World's most expensive spice. Available as threads or ground. | Check 'best before' date. | In an airtight container in a cool, dark, dry place. Use by recommended date. | Not applicable. |
| **Salt**<br>Available as rock salt, mined from the ground, or sea salt, extracted from sea water, as crystals, in a block, as flakes or fine granules. | Avoid damaged packaging. | In an airtight container in a cool, dark, dry place. Dampness causes lumps to form. Stored correctly, it keeps indefinitely. | Freeze in dishes. |
| **Sesame seeds**<br>These cream-coloured seeds, from a tropical tree, have a nutty aroma and flavour. | Check 'best before' date. Available whole or ground, white or toasted. | As poppy seeds. | Freeze in dishes. |

| Preparing and cooking | Typical uses | Alternative forms |
|---|---|---|
| Grind whole peppercorns as needed. Add at the end of cooking time. Complements all savoury dishes. | Salad dressings, peppered steak, marinades, sauces. | Fresh and pickled, freeze-dried, mustard, preserved in brine, pink peppercorns. |
| Use in sweet or savoury dishes. | Biscuits, cakes, cream-based dressings, curries, decoration on breads and rolls, garnish, pastries, sauces. | Paste, poppy-seed oil. |
| Quickly loses its flavour. Use for its beautiful yellow colour more than for its flavour. Colour is very strong, so only a pinch is needed. | Buns, cakes, chicken and fish dishes, curries, rice dishes such as paella. | |
| | Used mainly as seasoning, in both sweet and savoury dishes: batters, breads, casseroles, stews and soups; also for salting vegetables prior to cooking, and sprinkled over vegetables such as aubergine or cucumber to draw out bitter juices or water. | Flavoured salts such as celery, garlic or onion. Low-sodium salt. |
| Flavour is enhanced by dry-roasting. | Biscuits, cakes, stir-fries and other Chinese dishes, coating foods, crackers, garnish on bread, halva, hummus, sprinkled into salads and other dishes, tahini. | Sesame-seed oil and paste. |

| Food | How to choose | Storage | Freezing |
|------|---------------|---------|----------|
| **Star anise**<br>The dried fruit from an evergreen tree, this is a key ingredient of five-spice powder and has a pungent aniseed smell. Available whole, broken, ground or as seeds. | Check 'best before' date. | As poppy seeds. | Freeze in dishes. |
| **Turmeric**<br>The dried underground stem of a plant, this has a gingery/peppery smell and a pungent earthy flavour and is often included in curry powder. It is available whole, in pieces or ground. | Check 'best before' date. | As allspice. | Freeze in dishes. |
| **Vanilla**<br>The seed pod from a climbing orchid, this is deep brown in colour with a slightly smoky sweet taste and smell. Available as whole or split pods. | Check 'best before' date. Fresh whole pods should be tough yet supple and have a good dark-brown colour. | In a glass jar with a tight-fitting lid in a cool, dark, dry place.<br>Stored correctly, will keep indefinitely. Use by recommended date. | Not applicable. |

| Preparing and cooking | Typical uses | Alternative forms |
|---|---|---|
| | Flavouring chicken, other poultry, pork and fish dishes, Oriental dishes, stir-fried vegetables and pumpkin. | |
| Adds a brilliant yellow colour to dishes. | Curries; added to mustard blends, pickles and chutneys such as piccalilli; poultry dishes, soups, vegetable dishes. | |
| Pods can be kept in sugar to flavour it for cooking. | In sweet or savoury dishes: cakes, fillings, fruit compotes, ice-cream, milk puddings such as rice pudding and custard, salads, savoury dishes made with fish and veal, sweet sauces, syrups. | Essences, extract, powdered. |

# FOOD POISONING

| Type | Sources | Symptoms | Onset time | Duration of illness |
|------|---------|----------|------------|---------------------|
| Campylobacter | Poultry, meat, milk, untreated water, shellfish, household pets | Diarrhoea, abdominal pain, rarely vomiting. Blood and mucus may be excreted | 1-10 days (2-5 days average) | 2-7 days. Up to 4 weeks for full recovery |
| Salmonella | Poultry, meat, untreated milk, raw or inadequately cooked egg products, infected food handlers, household pets | General tiredness, diarrhoea, high fever, vomiting, abdominal pain. Blood poisoning (septicaemia) and inflammation of the abdominal wall (peritonitis) may develop in severe cases | 12-48 hours | Up to 3 weeks. Patient may be carrier for 12 weeks or longer after symptoms have gone |
| Listeria | Dairy products, meat, natural occurrence in environment, also commonly carried in human gut. Unpasteurised milk products and meat-based products have been cited specifically | From mild 'flu to septicaemia and meningitis. May cause miscarriage, stillbirth or premature labour in pregnant women | Varies. Outbreaks have occurred 3-70 days after food eaten | Varies |
| VTEC | Farm animals, undercooked minced beef including beefburgers, untreated cows' milk and cheese, contaminated pasteurised milk, infected people, untreated contaminated water | Abdominal cramps, vomiting, mild to severe diarrhoea that may contain blood. Children under 5 and some elderly people sometimes develop kidney problems (potentially but rarely fatal) | 1-8 days. Typically 2-4 days | Varies |

| Type | Sources | Symptoms | Onset time | Duration of illness |
|------|---------|----------|-----------|---------------------|
| *Staphylococcus aureus* | Human contamination of food. Causes skin and wound infections, carried in noses of 40% of healthy people and also found in gut | Vomiting, abdominal cramps, diarrhoea | 2-6 hours | Under 12 hours to 2 days |
| *Bacillus cereus* | Cereal products, spices, milk and dairy products, contaminated cooked foods (especially rice and pasta dishes) | Nausea, vomiting and stomach cramps, sometimes diarrhoea | 1-16 hours | Approx. 24-36 hours |
| *Clostridium perfringens* | Contaminated bulk-cooked meat and poultry dishes left at ambient temperature during cooling and storage | Diarrhoea, acute abdominal pain; vomiting uncommon | 8-18 hours | Approx. 24 hours |
| Small round structured viruses (SRSVs) | Humans. Spread through foods eaten uncooked or handled after cooking, shellfish and irrigated crops contaminated by human sewage | Nausea, vomiting (sometimes violent), diarrhoea | Approx. 12-48 hours | 2 days or less |

## What to do in the event of food poisoning

If you think you have food poisoning, see your GP. Doctors are required by law to report all cases of suspected food poisoning so that the local Environmental Health Department can investigate if necessary.

Your GP may ask you for a stool sample to identify the organism causing your illness. If you still have the suspect food, keep it in a sealed container in the refrigerator: the EHO may decide to send it for testing to see if the food and the stool contain similar organisms. Let your GP and your employer know if you think there is a risk of your transmitting the infection to others through your work (especially if you work with food or look after potentially vulnerable people such as the young, the old or the infirm).

With some types of food poisoning, such as VTEC or salmonella, you may feel well but the bacteria can stay in your gut long after the symptoms have gone – that is, you could be a 'carrier'. All the more reason to take great care with personal hygiene.

# ADDRESSES

Advertising Standards Authority, 2-16 Torrington Place, London WC1E 7HW
Tel. 0171-580 5555

The Department of Health
Tel. (Public Enquiry Office): 0171-210 4850.

Food Safety Advisory Centre produces information on food health and safety. Write to Foodline, 72 Rochester Row, London SW1P 1JU or phone free on 0800 282407.

The Guild of Food Writers, 48 Crabtree Lane, London SW6 6LW
Tel. 0171-610 1180
For copy of *Cooking in Metric* leaflet send A5 s.a.e.

Ministry of Agriculture, Fisheries and Food (MAFF) produces a range of *Foodsense* booklets on different food-related matters. Write to Foodsense, London SE99 7TT or phone MAFF on 0645 335577.

# ACKNOWLEDGEMENTS

The author and publishers would like to thank the following people and organisations for their assistance in providing information for this book:

Barrow Boar, Sarah Bradford, British Chicken Information Service, British Egg Information Service, British Turkey & Poussin Information Service, Canned Food Information Centre, Cauldron Foods, Food Safety Advisory Centre for information on page 27, Fresh Fruit and Vegetable Information Bureau, Frozen Food Information Bureau, The Guild of Food Writers, Marlow Foods, Ministry of Agriculture, Fisheries and Foods (MAFF) for reference information relating to pages 22-3, Meat and Livestock Commission, National Dairy Council, Rice Bureau, Schwartz Spices, Sea Fish Industry Authority, St Ivel Ltd, Tracey Tuffrey, Whitworths and Robert Williams.

The appendix on food poisoning is from *Health Which?*, April 1996. *Health Which?* is published six times a year by Which? Limited, the trading arm of Consumers' Association (a registered charity). To find out more about this or any other *Which?* magazine, write to Which? at PO Box 44, Hertford X, SG14 1SH.

# INDEX

# Which? Medicine

Have you ever been frustrated by the lack of information
on your medicine package? Do you feel that you don't get
enough advice from your doctor or pharmacist? *Which?
Medicine* explains in plain English all you need to know, and
particular attention is given to the use of medicines by
people over 65. It is the only wholly independent source of
drug information written exclusively for medicine users.

*Which? Medicine* takes a critical look at commonly used
prescription and non-prescription medicines and deals
with patients' unanswered questions. Introductory chapters
examine how medicines work and why people react in
different ways to the same preparation. The subsequent
sections go into more detail about particular body systems,
the common conditions and diseases and the drugs used to
combat them; they also explain what the wanted and
unwanted effects are.

Medicine record sheets have been devised to help you
monitor your use of medicines and communicate
effectively with your doctor or pharmacist – the first vital
step to being a full partner in the management of your and
your family's health care.

Paperback          210x120mm          512 pages

Available from bookshops, and by post from
Which?, Dept TAZM, Castlemead,
Gascoyne Way, Hertford X, SG14 1LH

You can also order using your credit card
by phoning FREE on (0800) 252100
(quoting Dept TAZM)